Springer Theses

Recognizing Outstanding Ph.D. Research

Aims and Scope

The series "Springer Theses" brings together a selection of the very best Ph.D. theses from around the world and across the physical sciences. Nominated and endorsed by two recognized specialists, each published volume has been selected for its scientific excellence and the high impact of its contents for the pertinent field of research. For greater accessibility to non-specialists, the published versions include an extended introduction, as well as a foreword by the student's supervisor explaining the special relevance of the work for the field. As a whole, the series will provide a valuable resource both for newcomers to the research fields described, and for other scientists seeking detailed background information on special questions. Finally, it provides an accredited documentation of the valuable contributions made by today's younger generation of scientists.

Theses may be nominated for publication in this series by heads of department at internationally leading universities or institutes and should fulfill all of the following criteria

- They must be written in good English.
- The topic should fall within the confines of Chemistry, Physics, Earth Sciences, Engineering and related interdisciplinary fields such as Materials, Nanoscience, Chemical Engineering, Complex Systems and Biophysics.
- The work reported in the thesis must represent a significant scientific advance.
- If the thesis includes previously published material, permission to reproduce this must be gained from the respective copyright holder (a maximum 30% of the thesis should be a verbatim reproduction from the author's previous publications).
- They must have been examined and passed during the 12 months prior to nomination.
- Each thesis should include a foreword by the supervisor outlining the significance of its content.
- The theses should have a clearly defined structure including an introduction accessible to new PhD students and scientists not expert in the relevant field.

Indexed by zbMATH.

Takaya Matsuura

Digital Quantum Information Processing with Continuous-Variable Systems

Doctoral Thesis accepted by
The University of Tokyo, Tokyo, Japan

Springer

Author
Dr. Takaya Matsuura
Department of Applied Physics
The University of Tokyo
Tokyo, Japan

Supervisor
Prof. Masato Koashi
Department of Applied Physics
Graduate School of Engineering
The University of Tokyo
Tokyo, Japan

Photon Science Center
Graduate School of Engineering
The University of Tokyo
Tokyo, Japan

ISSN 2190-5053 ISSN 2190-5061 (electronic)
Springer Theses
ISBN 978-981-19-8287-3 ISBN 978-981-19-8288-0 (eBook)
https://doi.org/10.1007/978-981-19-8288-0

This Springer imprint is published by the registered company Springer Nature Singapore Pte Ltd.
The registered company address is: 152 Beach Road, #21-01/04 Gateway East, Singapore 189721, Singapore

Supervisor's Foreword

Since the revelation that the unique properties of quantum mechanics can open up new possibilities in processing of information such as communication and computation, the quantum information processing (QIP) has been the subject of intensive research for more than a couple of decades. There are two branches of the QIP depending on how we may tailor physical systems to make them suitable for processing of information. One is discrete-variable (DV) QIP, which accesses physical systems through observables taking discrete values such as energies, particle numbers and spins. The other is continuous-variable (CV) QIP based on continuous-valued observables such as positions, momentums, and quadrature components in optical phase space. The CV QIP is particularly attractive in optics since detectors with excellent selectivity and nearly ideal efficiency are available at lower cost. On the other hand, the CV QIP requires much more efforts in developing a supporting theory due to complexity of mathematical tools to handle continuous variables and functions as well as necessity of eventually converting those to discrete ones for implementation with digital signal processing.

The present thesis work by Takaya Matsuura develops theoretical methods to solve many important problems in CV QIP for communication and computation, which are obtained by careful investigation on the connection between DV QIP and CV QIP. For quantum communication, it provides a long-sought full security proof for CV quantum key distribution (QKD). Security against arbitrary attacks for the CV QKD under a realistic condition such as finite communication rounds and the use of digitized information processing was a long-standing open problem. In the thesis, a binary-phase-shift-keying (BPSK)-type CV-QKD protocol is proposed and its finite-size security against general attacks is proved. For general QIP, the thesis provides sophistication of the theory for Gottesman-Kitaev-Preskill (GKP) codes, which are the most promising method for simulating DV QIP protocols on a physical system equipped for CV QIP. The ideal GKP-encoded state is unphysical and the use of its approximated versions is mandatory in practice. The thesis develops a unified theory to establish equivalence relations among conventionally used approximate GKP codes, which enables direct comparison between studies based on different approximations. Finally, for quantum computation, the thesis proposes a resource-efficient method to realize universal optical quantum computation using the GKP code. It is based on directly preparing a single GKP magic state instead of GKP Pauli states, which greatly simplifies existing protocols and reduces the required amount of resource states. Not to mention the importance of each of the above achievements, I believe that various technics and insights which helped in bringing about those in the entire thesis are useful for deeper understanding of bridge between CV QIP and DV QIP, which will benefit future development of QIP.

Tokyo, Japan
March 2021

Prof. Masato Koashi, Ph.D.

Preface

Continuous-variable quantum systems are characterized by physical quantities that can take continuous values. It appears in many physical systems, such as the position and momentum of oscillator modes of trapped ions and the complex amplitude of an optical mode. Especially, the quantum optical system is one of the most important continuous-variable systems for the quantum information application since light is an information carrier for telecommunication and behaves quantum mechanically even in the room-temperature environment. Furthermore, the complex amplitude of the optical mode can easily be manipulated. However, using a continuous variable, as it is, in quantum information processing does not work since a continuous variable is subject to continuous noises in the physical world, and such noises cannot completely be corrected. In order to perform reliable and fault-tolerant quantum information processing, information should be encoded in a digitized degree of freedom. In this thesis, digital quantum information processing with continuous-variables systems is studied in the quantum key distribution (QKD) and the quantum computation (QC), two major fields of quantum information processing.

For the QKD, the so-called continuous-variable QKD protocols, in which information is encoded on the complex amplitude of an optical pulse and the homodyne or heterodyne detector is used to measure it, are studied. More specifically, a continuous-variable QKD protocol that uses discretely modulated signals and the classical digital information processing is developed, and its security proof against general attacks in the finite-key case is established. This is achieved by virtually introducing the discrete-variable quantum system into the continuous-variable system and thereby reducing its security argument to that of discrete-variable QKDs, which is more mature. This is in sharp contrast to the previous security analyses on this kind of QKD protocol, where they directly use continuous variables in the security analyses and thus have difficulty in extending the analyses to the finite-key case. Another key to our security proof is the development of the method of estimating a lower bound on the fidelity to the coherent states, which eventually leads to the evaluation of how well the discrete-variable system can be approximated by the continuous-variable one. These security analyses are further elaborated, and a protocol whose key rate against the loss noise achieves almost optimal scaling is obtained.

For the QC with the continuous-variable system, the Gottesman-Kitaev-Preskill (GKP) code, which encodes discrete quantum information into a continuous-variable system, is studied. This code has many desirable properties for the optical implementation of QC; e.g., the universal QC is possible using only the Gaussian operations, which are relatively easy to implement in the quantum optical system, along with the GKP-encoded states. Because the ideal GKP-encoded state is unphysical, only its approximation in some sense is realizable. Thus, several approximations of the GKP-encoded state have been developed since the first proposal of the GKP code. In this thesis, these conventionally used approximations are shown to be equivalent, and the explicit correspondences between the approximation parameters are given. This enables the direct comparison between the previous studies that were based on the different approximations. As the final result, an efficient method to implement the universal fault-tolerant QC using the Gaussian operations and only one type of the GKP-encoded state is developed. Contrary to the previous method of using the probabilistic state conversion and the costly distillation, our method can prepare the necessary elements deterministically. Based on the previous proposals of generating approximate GKP-encoded states, the physical systems suitable for our method of realizing the universal QC are investigated. The results in this thesis thus broaden the possibilities for reliable quantum information processing with continuous-variable systems.

Tokyo, Japan Dr. Takaya Matsuura

List of Publications

Parts of this thesis have been published in the following journal articles:

- Takaya Matsuura, Kento Maeda, Toshihiko Sasaki, and Masato Koashi, "Finite-size security of continuous-variable quantum key distribution with digital signal processing," Nature Communications **12**, 252 (2021).
- Hayata Yamasaki, Takaya Matsuura, and Masato Koashi, "Cost-reduced all-Gaussian universality with the Gottesman-Kitaev-Preskill code: Resource-theoretic approach to cost analysis," Physical Review Research **2**, 023270 (2020).
- Takaya Matsuura, Hayata Yamasaki, and Masato Koashi, "Equivalence of Approximate Gottesman-Kitaev-Preskill codes," Physical Review A **102**, 032408 (2020).

Acknowledgements

I have received much support and advice during my Ph.D. studies. First, I would like to express my sincere thanks to my supervisor Masato Koashi for the instruction and all the constructive suggestions. He was always ready to listen to my questions and concerns for both academic and non-academic matters. From him, I learned the approaches to challenging problems and the importance of thorough consideration. I am grateful to all the collaborators, Kento Maeda, Toshihiko Sasaki, and Hayata Yamasaki, for the extensive discussions, which gave me fruitful ideas. Without their help, I could not complete the works presented in this thesis. I thank all the other group members, Yoshifumi Nakata, Wang Liao, Jingfang Zhou, Masanori Sumizaki, Shiro Tamiya, Shinichiro Yamano, Rei Tokami, and the former group members, Yui Kuramochi, Hongyi Zhou, Kenji Nishikawa, Khodai Kuroiwa, for valuable time having seminars and talking about scientific as well as everyday affairs. I especially thank Nakata-san for revising my application forms as well as presentations very carefully and Kuramochi-san for teaching me math.

I am grateful to Nicolas Menicucci and his group members, especially Ben Baragiola, for having accepted my visit to RMIT. It was my first time visiting the lab. overseas. I thank Christian Weedbrook and Ish Dhand for having given me an opportunity to work with people in Xanadu. I had fruitful discussions with Ish, Nick, Krishna Kumar Sabapathy, Rafael Alexander, Ilan Tzitrin, and all the other project members, which produced many results.

I have been financially supported by the ALPS program and the JSPS Research Fellowship for Young Scientists (DC2), which made my student life productive. I also thank my vice supervisor Akira Furusawa of the ALPS program for having given me lots of useful advice and put me in touch with Nick.

Many thanks to my parents for keeping me healthy in body and mind. They have been supporting me continuously since my birth. And finally, many thanks to my beloved, who is always with me and cheers me up. It is the future with her that makes me write this now.

Contents

Chapter 1
Introduction

Abstract In this chapter, general background for the research as well as what is treated in the thesis is summarized. Later, the structure of the thesis will be explained.

Keywords Continuous variable system · Quantum key distribution · Quantum computation

1.1 Background of the Research

Quantum information processing is an emerging technology having distinct advantages over ordinary information processing that prevails in the real world (referred to as classical information processing). Quantum key distribution (QKD) [1], as an example, enables distant parties to share the keys that are secret to the adversary even when the adversary has unlimited computational power and arbitrary eavesdropping technology allowed in the law of physics. This type of strong security cannot be obtained by classical information processing in a similar setup. Quantum computation (QC) [2–4], as another example, can efficiently solve problems that are considered to be difficult to solve in the classical computer [5]. Not limited to a matter of new technology, quantum information processing reveals plenty of essential aspects of quantum theory such as Bell non-locality [6–8] and entanglement, promoting our understanding of quantum theory. For these reasons, quantum information processing gathers growing interests and attention, and intensive studies have been made in this field for decades. Recently, the real-world implementation of quantum information technologies has been progressing. There are varieties of candidate materials and physical systems competing for realizing quantum information processing. Quantum optical system is one of the leading candidates for implementing quantum communication and quantum computation due to its retention ability of the quantum nature even in the room-temperature environment and its mobility with the propagating speed of light. Furthermore, technologies to control the wavy nature of light are established in the field of classical information processing.

T. Matsuura, *Digital Quantum Information Processing with Continuous-Variable Systems*, Springer Theses, https://doi.org/10.1007/978-981-19-8288-0_1

The complex amplitude of the optical mode takes continuous values and is thus called a continuous-variable quantum system. The relation between quantum information processing with the continuous-variable system and that with the discrete-variable one (qubit or qudit) is similar to the relation between analog and digital classical information processing. Continuous-variable quantum information processing is fragile against noises compared to the discrete-variable one, as analog classical information processing is compared to the digital one. To circumvent this issue, we need to "digitize" information on the continuous-variable quantum system and leave an irrelevant continuous degree of freedom as a redundancy. The idea is general and natural in principle, but it is non-trivial how to consistently encode digital information on the continuous-variable system and perform information processing on the encoded digital information with operations on the continuous-variable system. This thesis aims at developing quantum information processing on the digitized information encoded on the continuous-variable system.

1.2 What is Studied in This Thesis

As explained, the aim of the thesis is to develop digital quantum information processing with the continuous-variable system. We treat two topics; continuous-variable QKD and QC. For continuous-variable QKD, we develop a protocol that is adapted to digital information processing and prove the finite-size composable security for the protocol against general attacks. The complete security proof for the discrete-modulation continuous-variable QKD has been an open problem for a decade; we solve this problem by a newly developed estimation technique for the quantum state using the continuous-variable measurement. Our security analysis can easily incorporate the digitization of the signal processing and thus the finite resolution of the experimental apparatuses due to the adaptation to digital information processing. The security proof is further refined, and the protocol with almost optimal key rate scaling against transmission distance under the pure-loss channel is obtained.

For continuous-variable QC, we study the specific quantum error-correcting code called Gottesman-Kitaev-Preskill (GKP) code [9]. The GKP code was developed to encode a qubit into a continuous-variable system, targeting the complex amplitude of an optical mode. It is suitable for optical QC for several reasons, one of which is that the important subset of the elementary gate operations on the GKP-encoded qubit can be realized by reliably implementable operations in the quantum optical experiment. The GKP code can be implemented on the physical system only approximately, and thus several conventional approximations for the GKP code have been developed. However, the correspondence between different approximations for the GKP code is still unclear. In this thesis, the conventional approximations for the GKP code are shown to be equivalent by the explicit correspondence between the approximation parameters. This enables the translation between researches based on the different approximations, which has been a problem for the analyses of the approximate GKP codes. Further properties of the approximate GKP codes are also shown. Another result is the development of a resource-efficient protocol to realize the universal QC with the GKP code. This protocol relies on direct preparation of only one type of the GKP-encoded state, the GKP magic state, and requires no probabilistic state conversion process. The feasibility for the direct preparation of the GKP magic state is discussed for several experimental proposals that aim to implement the GKP-encoded state.

1.3 The Organization of This Thesis

In Chap. 2, the basics of quantum information theory that this thesis is based on are listed (Sect. 2.1), and the distance measures for quantum states used throughout the thesis are defined (Sect. 2.2). In Chap. 3, the definitions and the properties of the continuous-variable quantum system are reviewed (Sect. 3.1) with the quantum optical system as a prominent example (Sect. 3.2). Chapters 4 and 5 are the results of this thesis. In Chap. 4, after the review of the proof approach for the QKD and of the relevant probability theories, the security proof for a continuous-variable QKD protocol with digital information processing is developed (Sect. 4.4). The security proof is further refined based on the reverse reconciliation, a frequently used technique in the continuous-variable QKD, which results in the improvement of the protocol performance (Sect. 4.5). In Chap. 5, after reviewing the GKP code (Sect. 5.2.2), the equivalence of the conventionally-used approximate GKP codes is proved (Sect. 5.3). Furthermore, the resource-efficient protocol to realize the universal QC using the GKP code and experimentally feasible quantum optical operations is constructed (Sect. 5.4). Finally, Chap. 6 concludes the thesis.

References

1. Bennett CH, Brassard G (1984) Quantum cryptography: public key distribu tion and coin tossing. In: Proceedings of IEEE international conference on computers, systems, and signal processing. Bangalore, India, p 175
2. Benioff P (1980) The computer as a physical system: A microscopic quantum me chanical Hamiltonian model of computers as represented by turing machines. J Stat Phys 22(5):563–591. issn: 1572-9613. https://doi.org/10.1007/BF01011339
3. Feynman RP (1982) Simulating physics with computers. Int J Theor Phys 21(6):467–488. issn: 1572-9575. https://doi.org/10.1007/BF02650179
4. Deutsch D, Penrose R (1985) Quantum theory, the Church-Turing principle and the universal quantum computer. Proc R Soc Lond Math Phys Sci 400(1818):97–117. https://doi.org/10.1098/rspa.1985.0070, https://www.royalsocietypublishing
5. Shor PW (1994) Algorithms for quantum computation: discrete logarithms and fac toring. Proceedings 35th annual symposium on foundations of computer science, pp 124–134. https://doi.org/10.1109/SFCS.1994.365700
6. Bell JS (1964) On the Einstein Podolsky Rosen paradox. Phys Physique Fizika 1:195–200. https://doi.org/10.1103/PhysicsPhysiqueFizika, https://link.aps.org/doi/10.1103/PhysicsPhysiqueFizika

7. Freedman SJ, Clauser JF (1972) Experimental test of local hidden variable theories. Phys Rev Lett 28:938–941. https://doi.org/10.1103/PhysRevLett.28.938, https://link.aps.org/doi/10.1103/PhysRevLett.28.938
8. Aspect A, Dalibard J, Roger G (1982) Experimental test of Bell's inequalities using time-varying analyzers. Phys Rev Lett 49:1804–1807. https://doi.org/10.1103/PhysRevLett.49.1804, https://link.aps.org/doi/10.1103/PhysRevLett.49.1804
9. Gottesman D, Kitaev A, Preskill J (2001) Encoding a qubit in an oscillator. Phys Rev A 64(1):012310

Chapter 2
Preliminaries

Abstract In this chapter, preliminaries for the quantum information theory are explained. Then, two important measures for the closeness of quantum states, which are frequently used in this thesis, are introduced.

Keywords Quantum state · Quantum operation/channel · Quantum measurement/instrument · Qubit · Trace distance · Quantum fidelity

2.1 The Basic Principles of Quantum Information Theory

2.1.1 Linear Operator and Quantum State

In what follows, let $\mathcal{H}$ be a (possibly separable infinite-dimensional) Hilbert space. An element of $\mathcal{H}$ is denoted by a ket vector $|\psi\rangle$. An element of the dual $\mathcal{H}^*$ of $\mathcal{H}$ is denoted by a bra vector $\langle\psi|$. The inner product between two elements $|\psi\rangle$ and $|\phi\rangle$ in $\mathcal{H}$ is denoted by $\langle\psi|\phi\rangle$. Let $\mathfrak{L}(\mathcal{H})$ be the set of linear operators on $\mathcal{H}$. The dagger operation $\dagger$ on $A \in \mathfrak{L}(\mathcal{H})$ is defined as follows. Let $\mathcal{D}$ be the set of vectors so that for $|\phi\rangle \in \mathcal{D}$, there exists $|\phi'\rangle \in \mathcal{H}$ such that $\langle\phi|A|\psi\rangle = \langle\phi'|\psi\rangle$ holds for all $|\psi\rangle \in \operatorname{dom} A$. Then, $A^\dagger$ is defined to satisfy $|\phi'\rangle = A^\dagger|\phi\rangle$ (i.e., $\mathcal{D} = \operatorname{dom} A^\dagger$). Such an $A^\dagger$ can be uniquely defined as long as $\operatorname{dom} A$ is dense in $\mathcal{H}$. An operator $A \in \mathfrak{L}(\mathcal{H})$ is called self-adjoint if $A = A^\dagger$. By definition, $\operatorname{dom} A = \operatorname{dom} A^\dagger$ in this case. A self-adjoint operator A is called positive and denoted by $A \geq 0$ if $\langle\psi|A|\psi\rangle \geq 0$ holds for all $|\psi\rangle \in \operatorname{dom} A$. An operator $A \in \mathfrak{L}(\mathcal{H})$ is called bounded if there exists a constant $M > 0$ such that $\|A|\psi\rangle\| \leq M\||\psi\rangle\|$ holds for all $|\psi\rangle \in \mathcal{H}$ (i.e., $\operatorname{dom} A = \mathcal{H}$). The set of bounded linear operators on $\mathcal{H}$ is denoted by $\mathfrak{B}(\mathcal{H})$. An operator $A \in \mathfrak{B}(\mathcal{H})$ is called isometry if $\|A|\psi\rangle\| = \||\psi\rangle\|$ for all $|\psi\rangle \in \mathcal{H}$, and called unitary if $A^\dagger A = AA^\dagger = I$, where I denotes an identity operator on $\mathcal{H}$. In quantum information theory, we frequently consider linear operators from $\mathcal{H}_1$ to $\mathcal{H}_2$, denoted by $\mathfrak{L}(\mathcal{H}_1, \mathcal{H}_2)$. The set $\mathfrak{B}(\mathcal{H}_1, \mathcal{H}_2)$

T. Matsuura, *Digital Quantum Information Processing with Continuous-Variable Systems*, Springer Theses, https://doi.org/10.1007/978-981-19-8288-0_2

of bounded operators, an isometry, and a unitary from $\mathcal{H}_1$ to $\mathcal{H}_2$ can be defined analogously.

For a complete orthonormal system (CONS) $\{|e_j\rangle\}_{j\in\mathbb{N}} \subset \mathcal{H}$, the trace Tr of $A \in \mathfrak{B}(\mathcal{H})$ is defined as $\mathrm{Tr}(A) = \sum_{j\in\mathbb{N}}\langle e_j|A|e_j\rangle$ if it does not diverge. A bounded operator A with the trace of $\sqrt{A^\dagger A}$ convergent is called trace-class, and the set of trace-class operators is denoted by $\mathfrak{T}(\mathcal{H})$. A density operator $\rho \in \mathfrak{T}(\mathcal{H})$ is a positive and trace normalized operator, i.e.,

$$\rho^\dagger = \rho, \quad \rho \geq 0, \quad \mathrm{Tr}\rho = 1. \tag{2.1}$$

The set of density operators is denoted by $\mathfrak{D}(\mathcal{H})$.

A quantum state on a quantum mechanical system S can be uniquely determined by a density operator $\rho_S \in \mathfrak{D}(\mathcal{H}_S)$ on a Hilbert space $\mathcal{H}_S$ [1]. In the following, therefore, a quantum state is identified with its density operator. A quantum state is called pure if the rank of its density operator ρ is equal to one, i.e. $\rho = |\psi\rangle\langle\psi|$ for a unit norm vector $|\psi\rangle \in \mathcal{H}$. The pure state is often denoted by the unit vector for brevity, e.g., $|\psi\rangle$ in the above case. A quantum state is referred to as mixed if it is not pure. A quantum state on a composite system of systems S_1 and S_2 are given by a density operator $\rho_{S_1S_2} \in \mathfrak{D}(\mathcal{H}_{S_1} \otimes \mathcal{H}_{S_2})$ on the tensor product $\mathcal{H}_{S_1} \otimes \mathcal{H}_{S_2}$ of Hilbert spaces. For any density operator $\rho_S \in \mathfrak{D}(\mathcal{H}_S)$, there exists a pure state $|\psi\rangle_{SR} \in \mathcal{H}_S \otimes \mathcal{H}_R$ with a reference system R such that $\mathrm{Tr}_R|\psi\rangle\langle\psi|_{SR} = \rho_S$, where Tr_R denotes the partial trace of the system R. $|\psi\rangle_{SR}$ is called the purification of the density operator ρ_S. $|\psi\rangle_{SR}$ is uniquely determined up to isometry [2, 3].

2.1.2 *Quantum Operation and Quantum Channel*

A quantum operation on a quantum state of the system S is defined as a completely-positive (CP) and trace-non-increasing (TNI) linear map (CPTNI map in short) $\mathcal{E} : \mathfrak{T}(\mathcal{H}_S) \to \mathfrak{T}(\mathcal{H}_S)$, which satisfies

$$^\forall\rho_{SE} \in \mathfrak{D}(\mathcal{H}_S \otimes \mathcal{H}_E), \quad \mathcal{E} \otimes \mathrm{Id}_E(\rho_{SE}) \geq 0, \qquad \text{(Completely-positive)}, \tag{2.2}$$

$$^\forall\rho_S \in \mathfrak{D}(\mathcal{H}_S), \quad \mathrm{Tr}\left[\mathcal{E}(\rho_S)\right] \leq 1, \qquad \text{(Trace-non-increasing)}, \tag{2.3}$$

where $\mathrm{Id}_E : A \mapsto A, \ ^\forall A \in \mathfrak{B}(\mathcal{H}_E)$ denotes the identity map. These conditions are imposed to ensure the probabilistic interpretation of quantum states. The important subset of the quantum operations is the quantum channels, which are described as completely-positive and trace-preserving (TP) maps (CPTP maps in short). For CPTP maps, the inequality condition (2.3) of the CPTNI map is replaced with the equality. Roughly speaking, the quantum operation allows post-selection of the events while the quantum channel is non-selective. In general, the Hilbert spaces on which the input and the output trace-class operators of the CPTNI map are defined can be different. For a linear map $\mathcal{E} : \mathfrak{T}(\mathcal{H}_1) \to \mathfrak{T}(\mathcal{H}_2)$, the adjoint map of $\mathcal{E}$ denoted by $\mathcal{E}^\dagger$ from $\mathfrak{B}(\mathcal{H}_2)$ to $\mathfrak{B}(\mathcal{H}_1)$ is defined to satisfy

$$\mathrm{Tr}_{\mathcal{H}_2}[\mathcal{E}(\rho)A] = \mathrm{Tr}_{\mathcal{H}_1}[\rho\,\mathcal{E}^\dagger(A)], \quad \forall\rho \in \mathfrak{T}(\mathcal{H}_1),\ \forall A \in \mathfrak{B}(\mathcal{H}_2). \tag{2.4}$$

The adjoint of a CP map is also CP, the adjoint $\mathcal{F}$ of a TNI map is contractive, i.e., $\mathcal{F}(I) \le I$, and the adjoint $\mathcal{F}$ of a TP map is unital, i.e., $\mathcal{F}(I) = I$. Thus, the adjoint of a CPTNI map (i.e., a quantum operation) is a contractive CP map, and the adjoint of a CPTP map (i.e., a quantum channel) is a unital CP map. For a unital CP map $\mathcal{F} : \mathfrak{B}(\mathcal{H}_2) \to \mathfrak{B}(\mathcal{H}_1)$, there exists a Hilbert space $\mathcal{K}$, an isometry $V \in \mathfrak{B}(\mathcal{H}_1, \mathcal{K})$, and a homomorphism $\pi : \mathfrak{B}(\mathcal{H}_2) \to \mathfrak{B}(\mathcal{K})$ such that $\mathcal{F}(A) = V^\dagger\pi(A)V$ for all $A \in \mathfrak{B}(\mathcal{H}_2)$ [4]. This is called Stinespring dilation of the map $\mathcal{F}$, and the triple $(\pi, V, \mathcal{K})$ is unique up to isometry. For the adjoint map $\mathcal{E}^\dagger : \mathfrak{B}(\mathcal{H}_2) \to \mathfrak{B}(\mathcal{H}_1)$ of a CPTP map $\mathcal{E} : \mathfrak{T}(\mathcal{H}_1) \to \mathfrak{T}(\mathcal{H}_2)$, the Stinespring dilation takes a particularly simple form; $\mathcal{K} = \mathcal{H}_2 \otimes \mathcal{H}_R$ (where $\mathcal{H}_R$ can be chosen to be $\dim\mathcal{H}_R \le \dim\mathcal{H}_1 \times \dim\mathcal{H}_2$ if $\mathcal{H}_1$ and $\mathcal{H}_2$ are finite dimensional [5]), and $\pi(A) = A \otimes I_R$. Equivalently, the CPTP map $\mathcal{E} : \mathfrak{T}(\mathcal{H}_1) \to \mathfrak{T}(\mathcal{H}_2)$ can be written as

$$\mathcal{E}(\rho) = \mathrm{Tr}_R[V\rho V^\dagger], \quad \forall\rho \in \mathfrak{T}(\mathcal{H}_1), \tag{2.5}$$

which is (also) called Stinespring dilation of the CPTP map $\mathcal{E}$.

2.1.3 Quantum Measurement and Quantum Instrument

For a Borel-measurable space (X, Σ), a map $E : \Sigma \to \mathfrak{B}(\mathcal{H})$ is called positive-operator-valued measure (POVM) if it satisfies

$$0 \le E(\chi) \le I, \qquad \forall\chi \in \Sigma, \tag{2.6}$$

$$E(\emptyset) = 0, \quad E(X) = I, \tag{2.7}$$

$$E\Big(\bigcup_{k\in\mathbb{N}} \chi_k\Big) = \sum_{k\in\mathbb{N}} E(\chi_k), \quad \forall\{\chi_k\}_{k\in\mathbb{N}} \subseteq \Sigma \text{ with } \chi_j \cap \chi_k = \emptyset \text{ for } j \ne k, \tag{2.8}$$

where the summation of the right-hand side of Eq. (2.8) converges σ-weakly, i.e., $\mathrm{Tr}[\rho\sum_{k=1}^{n} E(\chi_k)] \xrightarrow{n\to\infty} \mathrm{Tr}[\rho\sum_{k\in\mathbb{N}} E(\chi_k)]$ for all $\rho \in \mathfrak{D}(\mathcal{H})$ [6, 7]. A quantum measurement with values in X is modeled by the above POVM if we do not care about the post-measurement state. Given a POVM $E : \Sigma \to \mathfrak{B}(\mathcal{H})$, the probability of observing $\chi \in \Sigma$ when the state of the system is prepared in ρ is given by

$$\Pr[\chi \in \Sigma|\rho] = \mathrm{Tr}[\rho E(\chi)], \qquad \text{(Born rule)}. \tag{2.9}$$

In other words, quantum measurement with values in X can be regarded as a (CPTP) map from the set of density operators to the set of probability distributions on X. For a POVM E, if $E(\chi)^2 = E(\chi)$ holds for all $\chi \in \Sigma$, then it is especially called projection-valued measure (PVM).

If we care about the post-measurement state, a quantum measurement should be treated as a quantum operation. Such quantum operations $\mathcal{E}_\chi$ $(\chi \in \Sigma)$ called the quantum instrument is a map from a σ-algebra Σ to the collection of CPTNI maps that satisfies

$$\mathcal{E}_\chi^\dagger(I) = E(\chi), \tag{2.10}$$

with an identity operator I on the post-measurement space and a POVM E [8].

For a measurement with finite number of outcomes $\{0, \dots, d-1\}$, one can associate the outcomes with mutually orthogonal pure states $\{|0\rangle\langle 0|, \dots, |d-1\rangle\langle d-1|\}$ on $\mathbb{C}^d$ called the classical states. The quantum channel $\mathcal{I}$ composed of the instrument $\mathcal{E}$ and the register C of the outcomes can be defined as

$$\mathcal{I}(\rho) = \sum_{i=0}^{d-1} |i\rangle\langle i|_C \otimes \mathcal{E}_i(\rho), \tag{2.11}$$

which is often used in combination with the quantum instrument.

2.1.4 *Qubit as an Information Unit*

Qubit is a quantum system characterized by the two-dimensional Hilbert space. It is an elemental unit of the quantum information theory in analogy with the bit in the classical information theory. Let $\{|0\rangle, |1\rangle\}$ be an orthonormal basis of a qubit. Throughout the thesis, the Pauli operators σ^X, σ^Y, and σ^Z are defined in this basis as

$$\sigma^X = \begin{bmatrix} 0 & 1 \\ 1 & 0 \end{bmatrix}, \quad \sigma^Y = \begin{bmatrix} 0 & -i \\ i & 0 \end{bmatrix}, \quad \sigma^Z = \begin{bmatrix} 1 & 0 \\ 0 & -1 \end{bmatrix}. \tag{2.12}$$

In this thesis, $\{|0\rangle, |1\rangle\}$ is called the Z basis of the qubit and $\{|0_X\rangle = |+\rangle = (|0\rangle + |1\rangle)/\sqrt{2}, |1_X\rangle = |-\rangle = (|0\rangle - |1\rangle)/\sqrt{2}\}$ is called the X basis of the qubit.

2.2 Measures for the Closeness of Quantum States

In this section, the definitions and properties of the trace distance and the fidelity are listed. The proofs of them are given, for example, in the book [9].

2.2.1 The Trace Norm and The Trace Distance

The trace distance is a distance measure of quantum states based on the trace norm defined below.

Definition 2.2.1 (*Trace norm*) The trace norm $\|\cdot\|_1$ of a trace-class operator $A \in \mathfrak{T}(\mathcal{H})$ is defined as

$$\|A\|_1 := \mathrm{Tr}\sqrt{A^\dagger A}. \tag{2.13}$$

As a norm, the trace norm satisfies the following properties.

- (Positive definiteness) The trace norm of a linear operator $A \in \mathfrak{T}(\mathcal{H})$ is positive semidefinite:

$$\|A\|_1 \geq 0. \tag{2.14}$$

  The trace norm is null if and only if the operator A is null:

$$\|A\|_1 = 0 \Leftrightarrow A = 0. \tag{2.15}$$

- (Homogeneity) For any constant $c \in \mathbb{C}$,

$$\|cA\|_1 = |c|\|A\|_1. \tag{2.16}$$

- (Subaddtivity) For $A, B \in \mathfrak{T}(\mathcal{H})$. Then the following inequality holds:

$$\|A + B\|_1 \leq \|A\|_1 + \|B\|_1. \tag{2.17}$$

The following property is given as a corollary.

Corollary 2.2.2 (Convexity) *For* $A, B \in \mathfrak{T}(\mathcal{H})$ *and* $\lambda \in [0, 1]$, *the following inequality holds:*

$$\|\lambda A + (1-\lambda)B\|_1 \leq \lambda\|A\|_1 + (1-\lambda)\|B\|_1. \tag{2.18}$$

The trace distance between quantum states ρ and σ is defined using the trace norm.

Definition 2.2.3 (*Trace distance*) The trace distance $d(\rho, \sigma)$ between two density operators ρ and σ is defined as

$$d(\rho, \sigma) = \frac{1}{2}\|\rho - \sigma\|_1 = \frac{1}{2}\mathrm{Tr}\sqrt{(\rho-\sigma)^\dagger(\rho-\sigma)}. \tag{2.19}$$

From the subadditivity property of the trace norm, $0 \leq d(\rho, \sigma) \leq 1$ holds. Furthermore, the following also holds.

Corollary 2.2.4 (Triangle inequality) *For $\rho, \sigma, \tau \in \mathfrak{D}(\mathcal{H})$, the following triangle inequality holds for trace distance:*

$$d(\rho, \sigma) \leq d(\rho, \tau) + d(\tau, \sigma). \tag{2.20}$$

The trace distance is a measure of how well two quantum states can be distinguished [9]. The trace distance satisfies the following desirable property in terms of information processing.

Proposition 2.2.5 (Monotonicity) *For any CPTP map $\mathcal{E} : \mathfrak{D}(\mathcal{H}) \to \mathfrak{D}(\mathcal{H})$ and for any quantum states $\rho, \sigma \in \mathfrak{D}(\mathcal{H})$, the following condition holds:*

$$d(\rho, \sigma) \geq d\left(\mathcal{E}(\rho), \mathcal{E}(\sigma)\right). \tag{2.21}$$

Monotonicity property of the trace distance reflects the fact that the distinguishability of two quantum states does not increase via any quantum channels. This property is crucial in the security proof of the QKD.

2.2.2 The Quantum Fidelity

The fidelity between two quantum states is defined as follows.

Definition 2.2.6 *(Fidelity)* Let ρ, σ be quantum states. The fidelity $F(\rho, \sigma)$ between ρ and σ is defined as

$$F(\rho, \sigma) := \left\| \sqrt{\rho}\sqrt{\sigma} \right\|_1^2 . \tag{2.22}$$

As a direct consequence, the following proposition holds:

Proposition 2.2.7 (Expected fidelity) *The fidelity between a pure state $|\psi\rangle$ and a mixed state ρ is given by*

$$F(|\psi\rangle, \rho) = \langle\psi|\rho|\psi\rangle. \tag{2.23}$$

The expected fidelity can be seen as the probability of observing "Yes" in the Yes/No measurement $\{|\psi\rangle\langle\psi|, I - |\psi\rangle\langle\psi|\}$ performed on the state ρ. One can also check that $0 \leq F(\rho, \sigma) \leq 1$. Although the fidelity is also a measure for the distinguishability of quantum states, one can notice that the distinguishability grows when the fidelity gets decreased, contrary to the trace distance. The followings are the useful properties of the fidelity.

Proposition 2.2.8 (Concavity) *Let ρ_1, ρ_2, σ be the quantum states and $\lambda \in [0, 1]$. Then the following inequality holds:*

$$F(\lambda\rho_1 + (1-\lambda)\rho_2, \sigma) \geq \lambda F(\rho_1, \sigma) + (1-\lambda)F(\rho_2, \sigma) \tag{2.24}$$

Proposition 2.2.9 (Monotonicity) *For any CPTP map $\mathcal{E} : \mathfrak{D}(\mathcal{H}) \to \mathfrak{D}(\mathcal{H})$ and for any quantum states $\rho, \sigma \in \mathfrak{D}(\mathcal{H})$, the following condition holds:*

$$F(\rho, \sigma) \leq F(\mathcal{E}(\rho), \mathcal{E}(\sigma)). \tag{2.25}$$

Theorem 2.2.10 (Uhlmann [10, 11]) *Let $\rho, \sigma \in \mathfrak{D}(\mathcal{H}_A)$ be quantum states on the system A, and R be the system with $\mathcal{H}_R \cong \mathcal{H}_A$. Then, the following expression holds for the fidelity:*

$$F(\rho, \sigma) = \max\{|\langle\psi|\phi\rangle|^2 : |\psi\rangle \in \mathcal{P}_\rho, |\phi\rangle \in \mathcal{P}_\sigma\}, \tag{2.26}$$

where $\mathcal{P}_\rho = \{|\psi\rangle \in \mathcal{H}_A \otimes \mathcal{H}_R : |\psi\rangle \text{is a purification of } \rho\}$.

The following lemma is not very common, so a proof is given here for later use.

Lemma 2.2.11 (Lower bound of the fidelity in the extended system) *For any quantum state $\tau_{AE} \in \mathfrak{D}(\mathcal{H}_A \otimes \mathcal{H}_E)$ and any pure state $|0\rangle\langle 0|_A \in \mathfrak{D}(\mathcal{H}_A)$,*

$$(F(\tau_A, |0\rangle\langle 0|_A))^2 \leq F(\tau_{AE}, |0\rangle\langle 0|_A \otimes \tau_E) \leq F(\tau_A, |0\rangle\langle 0|_A) \tag{2.27}$$

holds, where $\tau_E := \mathrm{Tr}_A \tau_{AE}$ and $\tau_A := \mathrm{Tr}_E \tau_{AE}$.

Proof The second inequality is the direct consequence of Lemma 2.2.9 applied on the partial trace Tr_E. In order ot prove the first inequality, let Q, R be auxiliary systems with $\mathcal{H}_R \cong \mathcal{H}_A$, and let $|\Psi\rangle_{AEQR}, |\Phi\rangle_{AEQR}$ be the purified states of τ_{AE} and $|0\rangle\langle 0|_A \otimes \tau_E$, respectively, which are given by

$$|\Psi\rangle_{AEQR} := \sum_i |i\rangle_A |\psi_i\rangle_{EQ} |\tilde{0}\rangle_R \tag{2.28}$$

$$|\Phi\rangle_{AEQR} := |0\rangle_A |\phi\rangle_{EQR}. \tag{2.29}$$

Here, $\{|i\rangle_A\}_i$ is a CONS in $\mathcal{H}_A$ that contains $|0\rangle_A$, $|\psi_i\rangle$s are subnormalized vectors satisfying $\sum_i \langle\psi_i|\psi_i\rangle = 1$, and $|\phi\rangle_{EQR}$ is given by

$$|\phi\rangle_{EQR} = \sum_i |\psi_i\rangle_{EQ} |\tilde{i}\rangle_R, \tag{2.30}$$

where $\{|\tilde{i}\rangle_R\}_i$ is a CONS in $\mathcal{H}_R$ that contains $|\tilde{0}\rangle_R$. One can ensure that $|\phi\rangle_{EQR}$ is a purification of τ_E because

$$\mathrm{Tr}_{QR}\left[|\phi\rangle\langle\phi|_{EQR}\right] = \sum_{i,j} \mathrm{Tr}_{QR}\left[|\psi_i\rangle\langle\psi_j|_{EQ} \otimes |\tilde{i}\rangle\langle\tilde{j}|_R\right] \tag{2.31}$$

$$= \sum_i \mathrm{Tr}_Q\left[|\psi_i\rangle\langle\psi_i|_{EQ}\right] \tag{2.32}$$

$$= \tau_E, \tag{2.33}$$

from the definition of τ_E and $|\psi_i\rangle$ in (2.28). By using Uhlmann's theorem (2.26), we have

$$F(\tau_{AE}, |0\rangle\langle 0|_A \otimes \tau_E) \geq \left|\langle\Psi|_{AEQR}|\Phi\rangle_{AEQR}\right|^2 \tag{2.34}$$
$$= \left|\langle\psi_0|_{EQ}\langle\tilde{0}|_R|\phi\rangle_{EQR}\right|^2 \tag{2.35}$$
$$= \left|\langle\psi_0|\psi_0\rangle\right|^2, \tag{2.36}$$

where we used Eq. (2.30) in the second equality. On the other hand, the following also holds:

$$F(\tau_A, |0\rangle\langle 0|_A) = \langle 0|_A \mathrm{Tr}_{EQR}\left[|\Psi\rangle\langle\Psi|_{AEQR}\right]|0\rangle_A \tag{2.37}$$
$$= \sum_{i,j}\langle 0|_A \mathrm{Tr}_{EQR}\left[|i\rangle\langle j|_A \otimes |\psi_i\rangle\langle\psi_j|_{EQ} \otimes |\tilde{0}\rangle\langle\tilde{0}|_R\right]|0\rangle_A \tag{2.38}$$
$$= \mathrm{Tr}_{EQ}|\psi_0\rangle\langle\psi_0|_{EQ} \tag{2.39}$$
$$= \langle\psi_0|\psi_0\rangle. \tag{2.40}$$

Combining these two, the following inequality holds:

$$(F(\tau_A, |0\rangle\langle 0|_A))^2 \leq F(\tau_{AE}, |0\rangle\langle 0|_A \otimes \tau_E). \tag{2.41}$$

Although the quantum fidelity itself is not a distance in mathematical sense, it has almost the same meaning as a distance since it is related to the trace distance as follows [1].

Proposition 2.2.12 (Relation between the trace distance and the fidelity) *Let ρ, σ be quantum states. Then,*

$$1 - \sqrt{F(\rho,\sigma)} \leq d(\rho,\sigma) \leq \sqrt{1 - F(\rho,\sigma)}. \tag{2.42}$$

The right-hand side of Eq. (2.42) is actually a distance measure called the sine distance or the purified distance [12–14].

References

1. Nielsen MA, Chuang IL (2010) Quantum computation and quantum information, 10th edn. Cambridge University Press, Cambridge. https://doi.org/10.1017/CBO9780511976667
2. Kirkpatrick KA (2006) The Schrödinger-HJW theorem. Found Phys Lett 19(1):95–102. issn: 1572-9524. https://doi.org/10.1007/s10702-006-1852-1, https://doi.org/10.1007/s10702-006-1852-1
3. Watrous J (2018) The theory of quantum information. Cambridge University Press, Cambridge
4. Forrest Stinespring W (1955) Positive functions on C*-algebras. Proc Am Math Soc 6(2):211–216

5. Choi M-D (1975) Completely positive linear maps on complex matrices. Linear Algebr Its Appl 10(3):285–290. issn: 0024-3795. https://doi.org/10.1016/0024-3795(75)90075-0, https://www.sciencedirect.com/science/article/pii/0024379575900750
6. Holevo AS (1973) Statistical decision theory for quantum systems. J Multivar Anal 3(4):337–394. issn: 0047-259X. https://doi.org/10.1016/0047-259X(73z90028-6), https://www.sciencedirect.com/science/article/pii/0047259X73900286
7. Holevo AS (2011) Probabilistic and statistical aspects of quantum theory, vol 1. Springer Science & Business Media
8. Davies EB, Lewis JT (1970) An operational approach to quantum probability. Commun Math Phys 17(3):239–260. issn: 1432-0916. https://doi.org/10.1007/BF01647093
9. Wilde MM (2013) Quantum information theory. Cambridge University Press, Cambridge
10. Uhlmann A (1976) The "transition probability" in the state space of a *-algebra. Rep Math Phys 9(2):273–279
11. Hou J, Qi X (2011) Fidelity of states in infinite dimensional quantum systems. arXiv: 1107.0354 [quant-ph]
12. Gilchrist A, Langford NK, Nielsen MA (2005) Distance measures to compare real and ideal quantum processes. Phys Rev A 71:062310. https://doi.org/10.1103/PhysRevA.71.062310
13. Rastegin AE (2006) Sine distance for quantum states. https://doi.org/10.48550/ARXIV.QUANT-PH/0602112, https://arxiv.org/abs/quant-ph/0602112
14. Tomamichel M (2012) A Framework for Non-Asymptotic Quantum Information Theory. PhD thesis. ETH Zurich

Chapter 3
Continuous-Variable Quantum System

Abstract Some physical systems, such as an oscillator mode of the trapped ion and a single optical mode, can be modeled by infinite-dimensional Hilbert spaces. Such a system has observables with a continuous spectrum, which are unique to the infinite dimension and can be used in quantum information processing [2, 3], and thus is called the continuous-variable quantum system.

Keywords Continuous-variable system · Canonical commutation relation · Gaussian operation · Quantum optics

3.1 Basics of the Continuous-Variable Quantum System

3.1.1 Position and Momentum Operators

Let $L^2(\mathbb{R})$ be the (normed) space of complex square-integrable functions on the real line $\mathbb{R}$. $L^2(\mathbb{R})$ is a Hilbert space with the inner product $\langle\psi|\phi\rangle = \int_{-\infty}^{\infty} dq\, \overline{\psi(q)}\phi(q)$. Let $X(s)$ and $Z(t)$ be (strongly continuous) one-parameter unitary groups on $L^2(\mathbb{R})$ defined by

$$[X(s)\psi](q) = \psi(q-s), \quad [Z(t)\psi](q) = e^{itq}\psi(q), \qquad \forall t, s \in \mathbb{R},\ \forall\psi \in L^2(\mathbb{R}), \tag{3.1}$$

and therefore satisfying $X(s)Z(t) = e^{-ist}Z(t)X(s)$. Due to the Stone's theorem, $X(s)$ and $Z(t)$ can be denoted by

$$X(s) = \exp(-is\hat{p}) \quad \text{and} \quad Z(t) = \exp(it\hat{q}) \tag{3.2}$$

with self-adjoint operators $\hat{q}$ and $\hat{p}$ satisfying

$$[\hat{q}\,\psi](q) = q\psi(q), \qquad \psi \in \operatorname{dom}\hat{q}, \tag{3.3}$$

$$[\hat{p}\,\phi](q) = i^{-1}\frac{d}{dq}\phi(q), \qquad \phi \in \operatorname{dom}\hat{p}. \tag{3.4}$$

T. Matsuura, *Digital Quantum Information Processing with Continuous-Variable Systems*, Springer Theses, https://doi.org/10.1007/978-981-19-8288-0_3

Self-adjoint operators $\hat{q}$ and $\hat{p}$ are called the position and the momentum operators, respectively, and can take continuous values. Written in terms of the Dirac's notation, they are

$$\hat{q} = \int_{-\infty}^{\infty} dq\, q|q\rangle\langle q|, \qquad \hat{p} = \int_{-\infty}^{\infty} dp\, p|p\rangle\langle p|, \tag{3.5}$$

$$X(s)^\dagger\, \hat{q}\, X(s) = \hat{q} + s, \qquad Z(t)^\dagger\, \hat{p}\, Z(t) = \hat{p} + t, \tag{3.6}$$

$$\langle q|q'\rangle = \delta(q - q'), \qquad \langle p|p'\rangle = \delta(p - p'), \qquad \langle q|p\rangle = \frac{1}{\sqrt{2\pi}} e^{iqp}, \tag{3.7}$$

$$\langle q|\psi\rangle = \psi(q), \qquad \langle p|\psi\rangle = \tilde{\psi}(p), \tag{3.8}$$

$$|\psi\rangle = \int_{-\infty}^{\infty} dq\, \psi(q)|q\rangle = \int_{-\infty}^{\infty} dp\, \tilde{\psi}(p)|p\rangle, \tag{3.9}$$

where $\delta(x)$ is the Dirac delta function and the $\tilde{\psi}$ is the Fourier transform of ψ. (Conventionally, $\psi(q)$ is called the position wave function and $\tilde{\psi}(p)$ is called the momentum wave function.) This particular Hilbert space and unitary groups $X(s)$ and $Z(t)$ on it play an important role, which is explained in the following. Let $U(s)$ and $W(t)$ $(s, t \in \mathbb{R})$ be one-parameter unitary groups satisfying $U(s)W(t) = e^{-ist}W(t)U(s)$. This is called Weyl-Segal canonical commutation relation (CCR). Equivalently, by introducing the vector $z = (s, t)$ and defining the unitary $V'(z) := e^{ist/2}U(s)W(t)$, the CCR can be denoted by $V'(z_1)V'(z_2) = \exp(i(t_1 s_2 - s_1 t_2)/2)\, V'(z_1 + z_2)$. The pair of the unitary groups $X(s)$ and $Z(t)$ in Eq. (3.1) is an irreducible representation of the CCR (called the Schrödinger representation). Likewise, the unitary $V(z)$ defined by

$$V(z) := e^{ist/2}X(s)Z(t) = e^{-ist/2}Z(t)X(s) = \exp\left(it\hat{q} - is\hat{p}\right) \tag{3.10}$$

is the Schrödinger representation of $V'(z)$. The Stone-von Neumann theorem states that any irreducible representation of CCR is unitarily equivalent to the Schrödinger representation. Therefore, we can always use the Schrödinger representation to analyze the continuous variables obeying the CCR. The more commonly known CCR is the commutation relation $[\hat{q}, \hat{p}] = i$ between the generators $\hat{q}$ and $\hat{p}$ (called Heisenberg CCR). Note that this commutation relation is meaningful only on a certain dense subset of the Hilbert space.

Now we turn to the multi-variable case. The Heisenberg CCR with N degrees of freedom can be written as $[\hat{q}_j, \hat{p}_k] = i\delta_{j,k}$, $[\hat{q}_j, \hat{q}_k] = 0$, and $[\hat{p}_j, \hat{p}_k] = 0$, where $\delta_{j,k}$ denotes the Kronecker delta. The corresponding Weyl form of the CCR is given as follows. Let $\boldsymbol{x} = (x_1, \ldots, x_N)$ and $\boldsymbol{y} = (y_1, \ldots, y_N)$ be N component row vectors. Define $X(\boldsymbol{x}) := \exp(-i\sum_{k=1}^{N} x_k \hat{p}_k)$ and $Z(\boldsymbol{y}) := \exp(i\sum_{k=1}^{N} y_k \hat{q}_k)$. Then, the Weyl CCR with N degrees of freedom is given by

$$X(\boldsymbol{x})Z(\boldsymbol{y}) = \exp\left(-i\sum_{k=1}^{N} x_k y_k\right) Z(\boldsymbol{y})X(\boldsymbol{x}). \tag{3.11}$$

For an alternative form of the above, let $z = (x_1, y_1, x_2, y_2, \ldots, x_N, y_N)$ be a $2N$ component vector, and let $\Delta(z, z')$ be a symplectic form defined as

$$\Delta(z, z') = \sum_{k=1}^{N} (x'_k y_k - x_k y'_k). \tag{3.12}$$

Then, an alternative form of the Weyl CCR with N degrees of freedom is given by

$$V(z)V(z') = e^{i\Delta(z,z')/2} V(z + z'). \tag{3.13}$$

The generator $\hat{R}(z)$ of $V(z)$ (i.e., $V(z) = e^{i\hat{R}(z)}$) can be given by

$$\hat{R}(z) = \sum_{k} (y_k \hat{q}_k - x_k \hat{p}_k), \tag{3.14}$$

and satisfies

$$[\hat{R}(z), \hat{R}(z')] = i\Delta(z, z')I. \tag{3.15}$$

3.1.2 Characteristic Function and Wigner Function

The quantum state in the continuous-variable system has convenient representations as a function of continuous variables. For $\rho \in \mathfrak{D}(\mathcal{H}^{\otimes N})$, its characteristic function $F_\rho(z)$ ($z \in \mathbb{R}^{2N}$) is defined as

$$F_\rho(z) = \mathrm{Tr}[\rho V(z)]. \tag{3.16}$$

The characteristic function $F_\rho(z)$ is continuous for all $z \in \mathbb{R}^{2N}$ and satisfies $|F_\rho(z)| \leq 1$, $F_\rho(\mathbf{0}) = 1$, and $F_\rho(-z) = (F_\rho(z))^*$ [1]. The Wigner function $W_\rho(\boldsymbol{r})$ ($\boldsymbol{r} \in \mathbb{R}^{2N}$) is defined as the (symplectic) Fourier transform of $F_\rho(z)$ given by [4]

$$W_\rho(\boldsymbol{r}) = \int_{-\infty}^{\infty} \frac{dz^{2N}}{(2\pi)^{2N}} F_\rho(z) \exp[-i\Delta(\boldsymbol{r}, z)]. \tag{3.17}$$

The characteristic function and the Wigner function has one-to-one correspondence to the density matrix. For the characteristic function, for example, the inverse transformation can be explicitly given by

$$\rho = \int_{-\infty}^{\infty} \frac{dz^{2N}}{(2\pi)^N} F_\rho(z) V(-z). \tag{3.18}$$

The Wigner function is real and satisfies a kind of normalization condition $\int d\boldsymbol{r}\, W_\rho(\boldsymbol{r}) = 1$. Thus, the Wigner function is called quasi-probability distribution; the only difference from the usual probability distribution is that the Wigner function can be negative while any probability distribution is nonnegative.

Similarly to probability distributions, the moment of the Wigner function can be defined. The first moment, the mean $\bar{\boldsymbol{r}}$, is defined as $\bar{\boldsymbol{r}} := \int d\boldsymbol{r}\, \boldsymbol{r} W_\rho(\boldsymbol{r})$ if it does not diverge. The second moment, the covariance matrix $\boldsymbol{V} = (V_{ij})$ is the $2N \times 2N$ matrix defined as

$$V_{ij} = \mathrm{Re}\left[\int d\boldsymbol{r}\, (r_i - \bar{r}_i)(r_j - \bar{r}_j) W_\rho(\boldsymbol{r})\right], \tag{3.19}$$

where $\mathrm{Re}[\cdot]$ denotes the real part. The higher moment can be defined similarly.

The Wigner function for ρ has the following alternative form:

$$W_\rho(\boldsymbol{r}) = \frac{1}{(2\pi)^N} \int_{-\infty}^{\infty} d\boldsymbol{x}\, e^{i\boldsymbol{p}\boldsymbol{x}^\top} \langle \boldsymbol{q} - \frac{\boldsymbol{x}}{2} | \rho | \boldsymbol{q} + \frac{\boldsymbol{x}}{2} \rangle, \tag{3.20}$$

where $\top$ denotes the transposition. This transformation from the density operator to the function over $\mathbb{R}^{2N}$ can be extended to an arbitrary operator A on $\mathcal{H}^{\otimes N}$, which is also called the Wigner function of A.

3.1.3 Gaussian States, Gaussian Channels, and Gaussian Measurements

The set of Gaussian states is a distinctive subset of the set of quantum states in continuous-variable systems; it is closed under the action of Gaussian channels, and a Gaussian state has lots of similarities to the classical probability distribution [5, 6].

A quantum state ρ in the continuous-variable system is called Gaussian if its characteristic function has a Gaussian form, i.e.,

$$F_\rho(z) = \exp\left(-\frac{1}{2} z \left(\boldsymbol{\Omega}^\top \boldsymbol{V} \boldsymbol{\Omega}\right) z^\top + i\bar{\boldsymbol{r}}\boldsymbol{\Omega} z^\top\right), \tag{3.21}$$

where $2N \times 2N$ matrix $\boldsymbol{\Omega}$ satisfies $\boldsymbol{x}\boldsymbol{\Omega}z^\top = \Delta(\boldsymbol{x}, z)$. One can show that its Wigner function is also Gaussian [6], i.e.,

$$W_\rho(\boldsymbol{r}) = \frac{\exp\left[-\frac{1}{2}(\boldsymbol{r}-\overline{\boldsymbol{r}})\boldsymbol{V}^{-1}(\boldsymbol{r}-\overline{\boldsymbol{r}})^\top\right]}{(2\pi)^N\sqrt{\det \boldsymbol{V}}}. \tag{3.22}$$

Examples of the Gaussian states are the vacuum state, coherent states and thermal states in the quantum optics. The covariance matrix $\boldsymbol{V}$ must satisfy a physicality condition $\boldsymbol{V} \geq \frac{i}{2}\boldsymbol{\Omega}$, which is reduced to the Heisenberg uncertainty principle $(\Delta q)^2(\Delta p)^2 \geq \frac{1}{4}$ in the case of $N = 1$ [3, 6–8].

A Gaussian unitary U is a unitary that transforms the position and momentum operators linearly. It is generated by self-adjoint operators that are at most quadratic in the position and momentum operators. The Weyl unitary $V(z)$ is contained in the set of Gaussian unitary operators since it adds a constant vector to the position and momentum operators. Other than the addition of a constant vector, the (adjoint) action of a Gaussian unitary U on the Weyl unitary $V(z)$ leads to $U^\dagger V(z)U = V(zS^{-1})$, where S denotes a symplectic matrix $S \in Sp(2N, \mathbb{R})$ that satisfies $\Delta(z, z') = \Delta(zS, z'S)$ or equivalently $S\boldsymbol{\Omega}S^\top = \boldsymbol{\Omega}$ [1, 8]. In other words, such unitary operators form a representation of $Sp(2N, \mathbb{R})$. Examples of the Gaussian unitaries are the phase rotations, squeezing, and beamsplitter coupling in quantum optics.

A Gaussian channel Φ is a CPTP map whose adjoint map $\Phi^\dagger$ acts on $V(z)$ as

$$\Phi^\dagger(V(z)) = V(zK)\exp\left[-\frac{1}{2}z\left(\boldsymbol{\Omega}^\top\Gamma\boldsymbol{\Omega}\right)z^\top + i\boldsymbol{m}\boldsymbol{\Omega}z^\top\right], \tag{3.23}$$

where K and Γ are $2N \times 2N$ real matrices and $\boldsymbol{m} \in \mathbb{R}^{2N}$ is a row vector [3, 8–11]. Therefore, the characteristic function of a density operator ρ transforms into

$$F_{\Phi(\rho)}(z) = F_\rho(zK)\exp\left[-\frac{1}{2}z\left(\boldsymbol{\Omega}^\top\Gamma\boldsymbol{\Omega}\right)z^\top + i\boldsymbol{m}\boldsymbol{\Omega}z^\top\right]. \tag{3.24}$$

For the Gaussian state ρ given in Eq. (3.21), the application of the Gaussian channel Φ on ρ leads to

$$\overline{\boldsymbol{r}} \to \overline{\boldsymbol{r}}K'^\top + \boldsymbol{m}, \tag{3.25}$$

$$\boldsymbol{V} \to K'\boldsymbol{V}K'^\top + \Gamma, \tag{3.26}$$

where K' denotes $K' = \boldsymbol{\Omega}K\boldsymbol{\Omega}^\top$. Therefore, a Gaussian state is transformed into a Gaussian state through the Gaussian channel. In fact, Eqs. (3.25) and (3.26) hold for general quantum states with the first and second moments given respectively by $\overline{\boldsymbol{r}}$ and $\boldsymbol{V}$. The complete positivity of the channel is equivalent to the following inequality for K' and Γ [8–11]:

$$\Gamma \geq \frac{i}{2}\boldsymbol{\Omega} - \frac{i}{2}K'\boldsymbol{\Omega}K'^\top. \tag{3.27}$$

A Gaussian channel is called quantum-limited when the equality in Eq. (3.27) is achieved [12, 13]. The Gaussian channel Φ can also be written as

$$\Phi(\rho) = \mathrm{Tr}_E\left[U(\rho \otimes \rho_E)U^\dagger\right], \tag{3.28}$$

where U is a Gaussian unitary and ρ_E is a Gaussian state [5, 8, 10, 14, 15]. Note that this is not the Stinespring dilation of the channel in general, since ρ_E can be a mixed state. If the state ρ_E is the vacuum state, the corresponding Gaussian channel is quantum-limited.

The classification of Gaussian channels in perspective of its information-theoretic characteristics has been intensively studied e.g. in Refs. [8–14, 16–29]. Among others, in the case of $N = 1$, the following channels are important in the application [10, 11, 18].

- The quantum-limited loss (pure-loss) channel: the Gaussian channel with $\boldsymbol{m}$, K, and Γ in Eqs. (3.25) and (3.26) being

$$\boldsymbol{m} = (0, 0), \quad K = \begin{pmatrix} \sqrt{\eta} & 0 \\ 0 & \sqrt{\eta} \end{pmatrix}, \quad \Gamma = \frac{1}{2}\begin{pmatrix} 1-\eta & 0 \\ 0 & 1-\eta \end{pmatrix}, \tag{3.29}$$

  where $0 \le \eta \le 1$. The parameter η is called the transmissivity of the loss channel.
- The quantum-limited amplifier channel: the Gaussian channel with $\boldsymbol{m}$, K, and Γ being

$$\boldsymbol{m} = (0, 0), \quad K = \begin{pmatrix} \sqrt{G} & 0 \\ 0 & \sqrt{G} \end{pmatrix}, \quad \Gamma = \frac{1}{2}\begin{pmatrix} G-1 & 0 \\ 0 & G-1 \end{pmatrix}, \tag{3.30}$$

  where $G \ge 1$. The parameter G is called the gain of the amplifier.
- The Gaussian random displacement channel: the Gaussian channel with $\boldsymbol{m}$, K, and Γ being

$$\boldsymbol{m} = (0, 0), \quad K = \begin{pmatrix} 1 & 0 \\ 0 & 1 \end{pmatrix}, \quad \Gamma = \begin{pmatrix} \sigma^2 & 0 \\ 0 & \sigma^2 \end{pmatrix}. \tag{3.31}$$

  The parameter σ^2 is called the variance of the Gaussian random displacement.

In fact, any phase-covariant single-mode Gaussian channel, that is, the channel in which the phase rotation transforms covariantly from the input to the output, can be realized by a composition of the quantum-limited loss and amplifier channels [10, 13]. For example, the Gaussian random displacement channel with the variance σ^2 can be realized by a quantum-limited loss channel with the transmissivity $\eta = 1/(1 + \sigma^2)$ followed by a quantum-limited amplifier channel with the gain $G = 1 + \sigma^2$.

Finally, we introduce the Gaussian measurement. Let $M : \mathcal{B}(\mathbb{R}^m) \to \mathfrak{B}(\mathcal{H}^{\otimes N})$ be a POVM on the continuous-variable system with N degrees of freedom, where $\mathcal{B}(\mathbb{R}^m)$ denotes the Borel σ-algebra of $\mathbb{R}^m$. The probability $\Pr_\rho[\cdot]$ that the outcome of the measurement on ρ is in $A \in \mathcal{B}(\mathbb{R}^m)$ is thus given by

$$\mathrm{Pr}_{\rho}[A] = \mathrm{Tr}\left[\rho \int_A M(d^m \xi)\right]. \tag{3.32}$$

Define the operator characteristic function $\phi_M(\boldsymbol{\omega})$ as follows:

$$\phi_M(\boldsymbol{\omega}) = \int e^{i\omega\xi^\top} M(d^m \xi), \qquad \boldsymbol{\omega} \in \mathbb{R}^m. \tag{3.33}$$

The POVM M is called Gaussian if its operator characteristic function $\phi_M(\boldsymbol{\omega})$ has the form [30, 31]

$$\phi_M(\boldsymbol{\omega}) = V(\boldsymbol{\omega} K \boldsymbol{\Omega}^\top) \exp\left(-\frac{1}{2}\boldsymbol{\omega}\boldsymbol{\alpha}\boldsymbol{\omega}^\top\right), \tag{3.34}$$

where K is a real $m \times 2N$ matrix and $\boldsymbol{\alpha}$ is a real symmetric $m \times m$ matrix satisfying

$$\boldsymbol{\alpha} \geq \pm \frac{i}{2} K \boldsymbol{\Omega} K^\top. \tag{3.35}$$

Let r_α be the rank of the matrix $\boldsymbol{\alpha}$ and $r_{\boldsymbol{\Omega}_K}$ be the rank of $K\boldsymbol{\Omega}K^\top$, which is always even. Then, the Gaussian POVM M on the system A is known to be written, with the continuous-variable system C with $r_\alpha - r_{\boldsymbol{\Omega}_K}/2$ degrees of freedom, in the following form [30, 31]:

$$M(d^m\xi) = \mathrm{Tr}_C\left[(I \otimes \rho_C)\, E_{AC}(d^m \xi)\right]. \tag{3.36}$$

Here, ρ_C is the centered (i.e., mean zero) Gaussian state with the covariance matrix $\boldsymbol{\alpha}_C$ satisfying

$$K P \Lambda \boldsymbol{\alpha}_C \Lambda P^\top K^\top = \boldsymbol{\alpha}, \tag{3.37}$$

where Λ changes the signs of the entries corresponding to the momentum variables and P denotes the projection onto the subspace spanned by $\mathrm{supp}(\boldsymbol{\alpha})K$ (see [30, 31] for more detail). The PVM E_{AC} is the spectral measure of the vector of self-adjoint operator:

$$\hat{X} = \hat{r}_A K^\top \otimes I_C + I_A \otimes \hat{r}_C \Lambda P^\top K^\top. \tag{3.38}$$

An example of the Gaussian measurement in the case of $N = 1$ is the PVM E that is the spectral measure of the position operator $\hat{q}$ given by

$$E(d\xi) = |(q =)\xi\rangle\langle(q =)\xi| d\xi, \tag{3.39}$$

where we followed the Dirac's notation in Eq. (3.5). It is the Gaussian measurement with $K = (1, 0)$ and $\boldsymbol{\alpha} = 0$ in its operator characteristic function (3.33) since

$$\phi_E(\omega) = \int e^{i\omega\xi} |(q =)\xi\rangle\langle(q =)\xi| d\xi = e^{i\omega\hat{q}} = V(0, \omega), \tag{3.40}$$

and

$$\boldsymbol{\Omega} = \begin{pmatrix} 0 & -1 \\ 1 & 0 \end{pmatrix}. \tag{3.41}$$

Such a single-outcome measurement with $\boldsymbol{\alpha} = 0$ is called the ideal homodyne measurement in the quantum optics [32]. A Gaussian measurement with $K = (1, 0)$ and $\alpha > 0$ can be regarded as a noisy version of the homodyne measurement.

Another example of the Gaussian measurement for $N = 1$ is the two-outcome measurement F given by

$$F(d^2\xi) = V(\sqrt{2}\boldsymbol{\xi})|0\rangle\langle 0|V(\sqrt{2}\boldsymbol{\xi})^\dagger \frac{d^2\xi}{\pi}, \tag{3.42}$$

where $|0\rangle$ denotes the vacuum state with its position wave function $\langle q|0\rangle$ given by

$$\langle q|0\rangle = \psi_0(q) = \pi^{-\frac{1}{4}} \exp(-q^2/2) \tag{3.43}$$

satisfying

$$[(\hat{q} + i\hat{p})\psi_0](q) = 0 \tag{3.44}$$

from Eqs. (3.3) and (3.4) and its Wigner function $W_{|0\rangle\langle 0|}(q, p)$ given by

$$W_{|0\rangle\langle 0|}(q, p) = \frac{1}{\pi} \exp(-q^2 - p^2). \tag{3.45}$$

The operator characteristic function (3.33) of the POVM F is given by

$$K = \frac{1}{\sqrt{2}} \begin{pmatrix} 1 & 0 \\ 0 & 1 \end{pmatrix}, \quad \boldsymbol{\alpha} = \frac{1}{4} \begin{pmatrix} 1 & 0 \\ 0 & 1 \end{pmatrix}. \tag{3.46}$$

This can be checked as follows. Let us observe that, for $s \in \mathbb{C}$,

$$V(-is, s)V(\sqrt{2}\xi_R, \sqrt{2}\xi_I)|0\rangle = e^{\sqrt{2}is(\xi_R + i\xi_I)} V(\sqrt{2}\xi_R, \sqrt{2}\xi_I)V(-is, s)|0\rangle \tag{3.47}$$

$$= e^{\sqrt{2}is(\xi_R + i\xi_I)} V(\sqrt{2}\xi_R, \sqrt{2}\xi_I)|0\rangle, \tag{3.48}$$

where the last equality follows from $V(-is, s) = \exp(is(\hat{q} + i\hat{p}))$ and Eq. (3.44). Changing s to $-s$ in the above, we also have

$$V(is, -s)V(\sqrt{2}\xi_R, \sqrt{2}\xi_I)|0\rangle = e^{-\sqrt{2}is(\xi_R + i\xi_I)} V(\sqrt{2}\xi_R, \sqrt{2}\xi_I)|0\rangle. \tag{3.49}$$

Combining these with Eq. (3.42), we have

$$V(-is, s)F(d^2\xi)V(is, -s)^\dagger = e^{\sqrt{2}is(\xi_R + i\xi_I) + \sqrt{2}is^*(\xi_R - i\xi_I)} F(d^2\xi), \tag{3.50}$$

where * denotes the complex conjugate. Integrating over $\mathbb{R}^2$, we have

$$V(-is, s)\, I\, V(is, -s)^\dagger = \phi_F\left(2\sqrt{2}\,\mathrm{Re}[s], -2\sqrt{2}\,\mathrm{Im}[s]\right). \quad (3.51)$$

Using Eq. (3.13), the left-hand side can be reformulated as

$$V(-is, s)\, I\, V(is, -s)^\dagger = V(-is, s)V(is^*, s^*) \quad (3.52)$$

$$= e^{i(is^*s-(-is)s^*)/2}V(-is + is^*, s + s^*) \quad (3.53)$$

$$= e^{-|s|^2}V(2\mathrm{Im}[s], 2\mathrm{Re}[s]). \quad (3.54)$$

Substituting $s = (\omega_R - i\omega_I)/(2\sqrt{2})$, we have

$$e^{-\frac{1}{8}(\omega_R^2+\omega_I^2)}V(-\omega_I/\sqrt{2}, \omega_R/\sqrt{2}) = \phi_F(\omega_R, \omega_I), \quad (3.55)$$

and thus Eq. (3.34) for $M = F$ holds with $\boldsymbol{\Omega}$, K, and $\boldsymbol{\alpha}$ given in Eqs. (3.41) and (3.46). Using Eq. (3.36), we can rewrite this POVM with the ancillary state ρ_C being the vacuum state $|0\rangle$ and the PVM E_{AC} being the joint spectral measure of $(\hat{q}_A + \hat{q}_C)/\sqrt{2}$ and $(\hat{p}_A - \hat{p}_C)/\sqrt{2}$. This can be checked by comparing Eqs. (3.37) and (3.38) with

$$P = \begin{pmatrix} 1 & 0 \\ 0 & 1 \end{pmatrix}, \quad \Lambda = \begin{pmatrix} 1 & 0 \\ 0 & -1 \end{pmatrix}, \quad (3.56)$$

and the covariance matrix $\boldsymbol{\alpha}_C$ of the vacuum being

$$\boldsymbol{\alpha}_C = \frac{1}{2}\begin{pmatrix} 1 & 0 \\ 0 & 1 \end{pmatrix} \quad (3.57)$$

from its Wigner function (3.45). This type of measurement is called the heterodyne (or dual-homodyne) measurement in the quantum optics [32]. Again, a Gaussian measurement with the same K but a larger value of $\boldsymbol{\alpha}$ can be interpreted as a noisy heterodyne measurement. The set of Gaussian measurements is essentially composed of multi-variable generalization of (noisy) symplectic-transformed homodyne and heterodyne measurements [30, 31].

3.1.4 Gaussian Operations

Gaussian operations are the important subset of operations that can be performed on the continuous-variable quantum system. It is composed of the preparation of Gaussian states, the action of Gaussian unitaries, Gaussian measurements, and the ignorance of a subsystem. Gaussian channels are contained in Gaussian operations, which is obvious from the expression (3.28). Note that in terms of the transformation law as in Eq. (3.23), the ignorance Φ_{tr} of a subsystem is given by $\Phi_{\mathrm{tr}}^\dagger(V(z)) = V(z \oplus (0, 0))$, where $\oplus$ denotes the direct sum. Implementation of Gaussian operations in quantum optical systems is relatively easy, which will be explained in the next section.

3.2 Quantum Optical System as a Continuous-Variable System

In this section, the quantum optical system is reviewed as a prominent example of the continuous-variable system. For detail, see e.g. [33, 34].

3.2.1 *Quantization of Electromagnetic Field*

The vector potential $\boldsymbol{A}(\boldsymbol{r}, t)$ of the free electromagnetic field satisfies the homogeneous wave equation

$$\nabla^2 \boldsymbol{A}(\boldsymbol{r}, t) - \frac{1}{c^2}\frac{\partial^2}{\partial t^2}\boldsymbol{A}(\boldsymbol{r}, t) = 0, \tag{3.58}$$

in the Coulomb gauge

$$\nabla \cdot \boldsymbol{A} = 0, \qquad \Phi = 0. \tag{3.59}$$

The electric field $\boldsymbol{E}(\boldsymbol{r}, t)$ and the magnetic field $\boldsymbol{B}(\boldsymbol{r}, t)$ are given by

$$\boldsymbol{E}(\boldsymbol{r}, t) = -\frac{\partial}{\partial t}\boldsymbol{A}(\boldsymbol{r}, t), \tag{3.60}$$

$$\boldsymbol{B}(\boldsymbol{r}, t) = \nabla \times \boldsymbol{A}(\boldsymbol{r}, t). \tag{3.61}$$

The Maxwell equation for electromagnetic fields follows from these equations. The solution of the wave equation (3.58) can be written as

$$\boldsymbol{A}(\boldsymbol{r}, t) = \sum_{\lambda}\int \frac{d\boldsymbol{k}}{\sqrt{\varepsilon_0}}\left[\boldsymbol{e}_\lambda(\boldsymbol{k})A_\lambda(\boldsymbol{k})u(\boldsymbol{k}; \boldsymbol{r}, t) + \boldsymbol{e}_\lambda^*(\boldsymbol{k})A_\lambda^*(\boldsymbol{k})u^*(\boldsymbol{k}; \boldsymbol{r}, t)\right], \tag{3.62}$$

where $A_\lambda(\boldsymbol{k})$ denotes the amplitude of the mode with the wave vector $\boldsymbol{k}$ and polarization λ, $\boldsymbol{e}_\lambda$ denotes the direction of the polarization, and $u(\boldsymbol{k}; \boldsymbol{r}, t)$ denotes the "normalized" solution of the wave equation (3.58), i.e.,

$$u(\boldsymbol{k}; \boldsymbol{r}, t) = \frac{e^{i\boldsymbol{k}\cdot\boldsymbol{r} - i\omega_{\boldsymbol{k}}t}}{\sqrt{(2\pi)^3 2\omega_{\boldsymbol{k}}}}, \tag{3.63}$$

with $\omega_{\boldsymbol{k}} = c|\boldsymbol{k}|$. Due to the condition of Coulomb gauge (3.59), we have $\boldsymbol{k} \cdot \boldsymbol{e}_\lambda(\boldsymbol{k}) = 0$. When the electromagnetic fields are quantized, the classical amplitude $A_\lambda(\boldsymbol{k})$ and its complex conjugate $A_\lambda^*(\boldsymbol{k})$ is replaced with $\sqrt{\hbar}$ times the operators $\hat{a}_\lambda(\boldsymbol{k})$ and $\hat{a}_\lambda^\dagger(\boldsymbol{k})$ that satisfies

$$[\hat{a}_\lambda(\boldsymbol{k}), \hat{a}_{\lambda'}^\dagger(\boldsymbol{k}')] = \delta_{\lambda,\lambda'}\delta^3(\boldsymbol{k} - \boldsymbol{k}'). \tag{3.64}$$

(These are not operators in the usual sense but rather the operator-valued distributions.) The resulting quantized vector potential as well as electromagnetic fields are given by

$$\hat{A}(r,t) = \sum_\lambda \int dk \sqrt{\frac{\hbar}{\varepsilon_0}} \left[e_\lambda(k)\hat{a}_\lambda(k)u(k;r,t) + e_\lambda^*(k)\hat{a}_\lambda^\dagger(k)u^*(k;r,t) \right], \tag{3.65}$$

$$\hat{E}(r,t) = \sum_\lambda \int dk\, i \sqrt{\frac{\hbar}{\varepsilon_0}} \left[e_\lambda(k)\hat{a}_\lambda(k)\omega_k u(k;r,t) - e_\lambda^*(k)\hat{a}_\lambda^\dagger(k)\omega_k u^*(k;r,t) \right], \tag{3.66}$$

$$\hat{B}(r,t) = \sum_\lambda \int dk\, i \sqrt{\frac{\hbar}{\varepsilon_0}} \left[(k \times e_\lambda(k))\hat{a}_\lambda(k)u(k;r,t) - (k \times e_\lambda^*(k))\hat{a}_\lambda^\dagger(k)u^*(k;r,t) \right]. \tag{3.67}$$

The energy H of the (quantized) electromagnetic field is given by

$$H = \frac{\varepsilon_0}{2} \int dr \left[\hat{E}^2(r,t) + c^2 \hat{B}^2(r,t) \right] \tag{3.68}$$

$$= \sum_\lambda \int dk\, \frac{\hbar\omega_k}{2} [\hat{a}_\lambda(k)\hat{a}_\lambda^\dagger(k) + \hat{a}_\lambda^\dagger(k)\hat{a}_\lambda(k)], \tag{3.69}$$

where we used $e_\lambda^*(k) \cdot e_{\lambda'}(k) = \delta_{\lambda,\lambda'}$ and

$$2 \int dr\, \omega_k u^*(k;r,t)u(k';r,t) = \delta^3(k - k'). \tag{3.70}$$

Next we construct a well-defined mode functions $v_{j,\lambda}(r,t)$. This can be made by taking an orthonormal system of weight functions $f_j(k)$, and define the mode function $v_{j,\lambda}(r,t)$ by

$$v_{j,\lambda}(r,t) = \int dk\, f_j^*(k) e_\lambda(k) \omega_k u(k;r,t) \tag{3.71}$$

and the mode annihilation operators $\hat{a}_{j,\lambda}$ by

$$\hat{a}_{j,\lambda} = \int dk\, f_j(k)\hat{a}_\lambda(k). \tag{3.72}$$

Due to the relation $\sum_j f_j^*(k) f_j(k') = \delta^3(k - k')$, we have

$$\hat{E}(r,t) = i \sqrt{\frac{\hbar}{\varepsilon_0}} \sum_\lambda \sum_j \left[v_{j,\lambda}(r,t)\hat{a}_{j,\lambda} - v_{j,\lambda}^*(r,t)\hat{a}_{j,\lambda}^\dagger \right]. \tag{3.73}$$

The electromagnetic field can thus be decomposed into the well-defined mode functions. By carefully designing the weight function $f_j(k)$ and the polarization vector

$\boldsymbol{e}_\lambda(\boldsymbol{k})$, we can obtain a pulse-shaped mode function. In this thesis, we assume that an optical pulse emitted by a light source can be treated as a single pulse-shaped mode. The mode annihilation and creation operators satisfy the commutation relation

$$[\hat{a}_{j,\lambda}, \hat{a}_{k,\lambda'}] = 0, \qquad [\hat{a}^\dagger_{j,\lambda}, \hat{a}^\dagger_{k,\lambda'}] = 0, \qquad [\hat{a}_{j,\lambda}, \hat{a}^\dagger_{k,\lambda'}] = \delta_{jk}\delta_{\lambda\lambda'}. \tag{3.74}$$

The electromagnetic field has the pure vacuum state $|0\rangle$ satisfying

$$\hat{a}_{j,\lambda}|0\rangle = 0, \quad \forall j, \lambda. \tag{3.75}$$

Since the electromagnetic field can be decomposed into the mode functions $\boldsymbol{v}_{j,\lambda}(\boldsymbol{r}, t)$, the vacuum state $|0\rangle$ can be interpreted an infinite tensor product of the vacua $|0\rangle_{j,\lambda}$ for each mode j and the polarization λ satisfying $\hat{a}_{j,\lambda}|0\rangle_{j,\lambda} = 0$. We can define the Fock state $|n\rangle_{j,\lambda}$ for the mode j and the polarization λ by

$$|n\rangle_{j,\lambda} = \frac{1}{\sqrt{n!}}(\hat{a}^\dagger_{j,\lambda})^n|0\rangle_{j,\lambda}, \tag{3.76}$$

that satisfies

$$\hat{n}_{j,\lambda}|n\rangle_{j,\lambda} = \hat{a}^\dagger_{j,\lambda}\hat{a}_{j,\lambda}|n\rangle_{j,\lambda} = n|n\rangle_{j,\lambda}, \tag{3.77}$$

where $\hat{n}_{j,\lambda} := \hat{a}^\dagger_{j,\lambda}\hat{a}_{j,\lambda}$ is called the photon number operator. The set of vectors $\{|n\rangle_{j,\lambda}\}$ forms a CONS of the Hilbert space associated with the mode j and the polarization λ. Finally, by setting $\hat{q}_{j,\lambda} = \frac{1}{\sqrt{2}}(\hat{a}_{j,\lambda} + \hat{a}^*_{j,\lambda})$ and $\hat{p}_{j,\lambda} = -i\frac{1}{\sqrt{2}}(\hat{a}_{j,\lambda} - \hat{a}^*_{j,\lambda})$, we have the commutation relation

$$[\hat{q}_{j,\lambda}, \hat{q}_{k,\lambda'}] = 0, \qquad [\hat{p}_{j,\lambda}, \hat{p}_{k,\lambda'}] = 0, \qquad [\hat{q}_{j,\lambda}, \hat{p}_{k,\lambda'}] = i\delta_{jk}\delta_{\lambda\lambda'}. \tag{3.78}$$

This is nothing but the multi-mode generalization of the Heisenberg CCR explained in the previous section. One can also check the consistency between $\hat{a}_{j,\lambda}|0\rangle_{j,\lambda} = 0$ and Eq. (3.44) for the vacuum state. All the equations given in this section are formal and not mathematically rigorous. More rigorous treatments are given in e.g. [35].

3.2.2 *Operations in Quantum Optics*

In the following, we suppress the indices for the mode and polarization degrees of freedom and add them only when needed. Linear optics is composed of optical media that responds linearly to the electric field. Linear optical unitaries are unitaries that can be realized by the combination of beamsplitter $B_{ij}(\theta)$ and the phase shifter $R(\phi)$ [36], given respectively by

$$B_{ij}(\theta) := \exp\left[-\theta(\hat{a}_i^\dagger \hat{a}_j - \hat{a}_j^\dagger \hat{a}_i)\right], \tag{3.79}$$

$$R(\phi) := \exp(i\phi \hat{a}^\dagger \hat{a}). \tag{3.80}$$

They respectively transform the annihilation operators into

$$B_{ij}(\theta)^\dagger \begin{pmatrix} \hat{a}_i \\ \hat{a}_j \end{pmatrix} B_{ij}(\theta) = \begin{pmatrix} \cos\theta & -\sin\theta \\ \sin\theta & \cos\theta \end{pmatrix} \begin{pmatrix} \hat{a}_i \\ \hat{a}_j \end{pmatrix}, \tag{3.81}$$

$$R(\phi)^\dagger \hat{a} R(\phi) = e^{i\phi}\hat{a}, \tag{3.82}$$

by the adjoint actions. (Note that $B_{ij}(\theta)^\dagger = B_{ij}(-\theta)$ as well as $R(\phi)^\dagger = R(-\phi)$ holds.) In particular, $R(\pi/2)$ corresponds to the Fourier transform between quadrature operators, i.e.,

$$R(-\pi/2)\hat{q}R(\pi/2) = -\hat{p}, \qquad R(-\pi/2)\hat{p}R(\pi/2) = \hat{q}. \tag{3.83}$$

For the beamsplitter B_{ij}, $\cos^2\theta$ and $\sin^2\theta$ are respectively called the transmissivity and the reflectivity. Contrary to linear optics, nonlinear optics can be realized by a nonlinear optical media whose polarization density responds nonlinearly to the electric field. An example of the (second-order) nonlinear optical process is optical squeezing. The squeezing unitary $S(\xi)$ ($\xi \in \mathbb{C}$) is given by

$$S(\xi) := \exp\left(\frac{1}{2}(\xi^* \hat{a}^2 - \xi \hat{a}^{\dagger 2})\right). \tag{3.84}$$

The combination of linear optical unitary (that contains phase space displacement) and squeezing unitary forms the Gaussian unitary operations introduced in the previous section.

An example of the optical single-mode Gaussian state is the coherent state $|\alpha\rangle$ defined as

$$|\alpha\rangle = e^{-\frac{|\alpha|^2}{2}} \sum_{n=0}^{\infty} \frac{\alpha^n}{\sqrt{n!}} |n\rangle. \tag{3.85}$$

It is known that the laser light is well described by the coherent state. Its position wave function is given by

$$\langle q|\alpha\rangle = \pi^{-\frac{1}{4}} \exp\left[-\frac{(q-\sqrt{2}\alpha_R)^2}{2} + iq\sqrt{2}\alpha_I - i\alpha_R\alpha_I\right], \tag{3.86}$$

where $\alpha = \alpha_R + i\alpha_I$. The displacement operator $D(\alpha)$ transforms the vacuum state into a coherent state with the amplitude $\alpha \in \mathbb{C}$, i.e.,

$$D(\alpha)|0\rangle = |\alpha\rangle. \tag{3.87}$$

In terms of the creation and annihilation operators, it is given by

$$D(\alpha) = \exp(\alpha \hat{a}^\dagger - \alpha^* \hat{a}). \tag{3.88}$$

The operator $V(z)$ introduced in the previous section is related to the displacement operator by

$$V(x, y) = D\left((x + iy)/\sqrt{2}\right). \tag{3.89}$$

Using this, the POVM F of the heterodyne measurement Eq. (3.42) can be rewritten by the coherent state vector $|\alpha\rangle$ as

$$F(d^2\alpha) = |\alpha\rangle\langle\alpha| \frac{d^2\alpha}{\pi}. \tag{3.90}$$

Another example of the Gaussian state is the squeezed state given by

$$S(\xi)|0\rangle = \frac{1}{\sqrt{\cosh r}} \sum_{n=0}^{\infty} \frac{\sqrt{(2n)!}}{2^n n!} \left(-e^{i\varphi} \tanh r\right)^n |2n\rangle, \tag{3.91}$$

where $\xi = re^{i\varphi}$. Written in terms of the position wave function, it is given by

$$\langle q|S(\xi)|0\rangle = \left(2\pi(\Delta q)^2\right)^{-\frac{1}{4}} \exp\left[-\frac{1 + i \sinh(2r) \sin\varphi}{4(\Delta q)^2} q^2\right], \tag{3.92}$$

where

$$(\Delta q)^2 := \frac{1}{2}\left[e^{2r} \sin^2(\varphi/2) + e^{-2r} \cos^2(\varphi/2)\right]. \tag{3.93}$$

A single-mode pure Gaussian state is in fact given by the squeezed coherent state $D(\alpha)S(\xi)|0\rangle$.

The homodyne detector is a detector that is frequently used in quantum optics and can be implemented by linear optical elements. The experimental setups of the homodyne detector are presented in Fig. 3.1. First, the optical signal is coupled with the local oscillator (that is, the strong coherent light used as a phase reference) by the 50:50 beamsplitter. The coupled modes are successively put into the photo-diodes. In photo-diodes, the photo-currents proportional to the intensity of the input electric field are generated. Since the intensity of the electric field is proportional to the photon number $\hat{n}$, the photo-currents I_1 and I_2 are proportional to

$$I_1 \propto \hat{n}_1 = \frac{1}{2}(\hat{a}_s^\dagger + \alpha^*)(\hat{a}_s + \alpha), \tag{3.94}$$

$$I_2 \propto \hat{n}_2 = \frac{1}{2}(\hat{a}_s^\dagger - \alpha^*)(\hat{a}_s - \alpha), \tag{3.95}$$

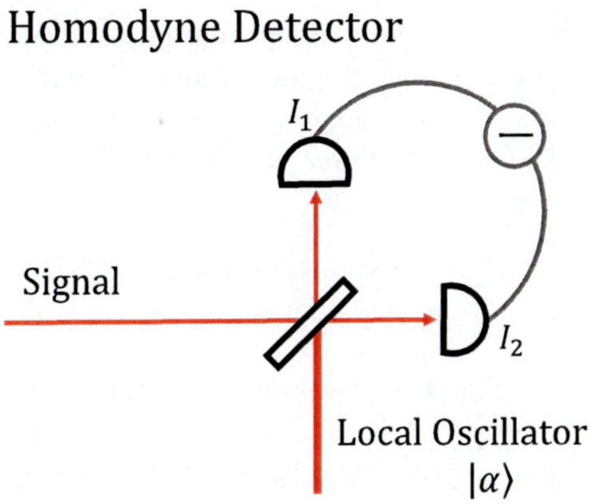

Fig. 3.1 The schematics of the homodyne detector. The signal light is coupled with the local oscillator $|\alpha\rangle$ by the 50:50 beamsplitter and successively measured by the photo-diodes. The difference $I_1 - I_2$ of the photo-currents is the output of the detector

where $\hat{a}_s$ denotes the annihilation operator of the signal mode. Taking the difference $I_1 - I_2$ of the photo currents, we have

$$I_1 - I_2 \propto |\alpha|(e^{-i\theta}\hat{a}_s + e^{i\theta}\hat{a}_s^\dagger) \tag{3.96}$$

$$\propto |\alpha|(\cos\theta\,\hat{q}_s + \sin\theta\,\hat{p}_s). \tag{3.97}$$

In the actual experiment, photo-diodes have electric noise, which makes the measurement carried out by the above setups unsharp. However, taking the intensity of the local oscillator infinitely strong, i.e., $|\alpha| \to \infty$, the homodyne measurement gets close to the quadrature measurement with its PVM given by $|q_\theta\rangle\langle q_\theta| dq_\theta$, where $\hat{q}_\theta := \cos\theta\,\hat{q}_s + \sin\theta\,\hat{p}_s$. (The case of $\theta = 0$ is exactly equal to the one in Eq. (3.39).)

Another detector that is frequently used in quantum optics is the heterodyne detector (also known as the dual-homodyne detector) depicted in Fig. 3.2. In this detector, the signal light is split into two by the 50:50 beamsplitter, and each of the two split modes is measured by the homodyne detector. One of the optical phases of the

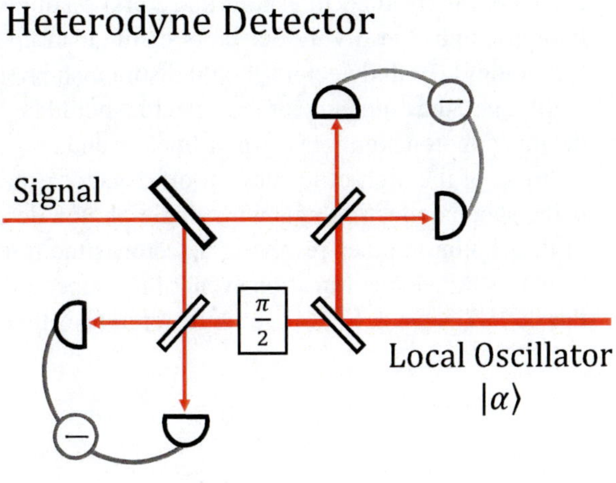

Fig. 3.2 The schematics of the heterodyne (dual-homodyne) detector. The signal light is split into two (i.e., is coupled with the vacuum) by the 50:50 beamsplitter and then measured by the homodyne detectors. One of the optical phases of the local oscillators is $\pi/2$-rotated, and thus two homodyne detectors measure orthogonal quadratures

local oscillators for the homodyne detectors is $\pi/2$-rotated so that the two homodyne detectors measure the orthogonal quadratures, e.g., $\hat{q}$ and $\hat{p}$. Precisely speaking, the first 50:50 beamsplitter couples the signal light and the external vacuum. Since the 50:50 beamsplitter transforms quadrature operators as

$$B_{12}(-\pi/4)^\dagger \hat{q}_1 B_{12}(-\pi/4) = (\hat{q}_1 + \hat{q}_2)/\sqrt{2}, \quad B_{12}(-\pi/4)^\dagger \hat{p}_2 B_{12}(-\pi/4) = (\hat{p}_1 - \hat{p}_2)/\sqrt{2}, \tag{3.98}$$

the 50:50 beamsplitter followed by the homodyne measurements for $\hat{q}_1$ and $\hat{p}_2$ in Fig. 3.2 is equivalent to the measurement for the joint spectral measure of $(\hat{q}_1 + \hat{q}_2)/\sqrt{2}$ and $(\hat{p}_1 - \hat{p}_2)/\sqrt{2}$. Combining these with Eq. (3.36), one can notice that the POVM for the detector in Fig. 3.2 is equal to the heterodyne POVM in Eq. (3.42) or equivalently (3.90) as explained in Sect. 3.1.3. An alternative derivation for this fact is as follows. The operator characteristic function of the two homodyne measurements at the end in Fig. 3.2 is given by $V(0, \omega_R, -\omega_I, 0)$. The transformation law for the operator $V(x_1, y_1, x_2, y_2)$ by the adjoint of the channel $\Phi_B(\rho) := B(-\pi/4)\rho B(-\pi/4)^\dagger$ is given as follows:

$$\Phi_B^\dagger(V(x_1, y_1, x_2, y_2)) = V\left(\frac{x_1 - x_2}{\sqrt{2}}, \frac{y_1 - y_2}{\sqrt{2}}, \frac{x_1 + x_2}{\sqrt{2}}, \frac{y_1 + y_2}{\sqrt{2}}\right). \tag{3.99}$$

Thus, the operator characteristic function before the first 50:50 beamsplitter that couples the signal light and the external vacuum $|0\rangle$ in Fig. 3.2 is given by

$$\begin{aligned} \langle 0| V\left(\frac{\omega_I}{\sqrt{2}}, \frac{\omega_R}{\sqrt{2}}\right)|0\rangle V\left(-\frac{\omega_I}{\sqrt{2}}, \frac{\omega_R}{\sqrt{2}}\right) &= \exp\left[-\frac{1}{8}(\omega_R^2 + \omega_I^2)\right] V\left(-\frac{\omega_I}{\sqrt{2}}, \frac{\omega_R}{\sqrt{2}}\right) \\ &= \phi_F(\omega_R, \omega_I), \end{aligned} \tag{3.100}$$

where $\phi_F(\omega_R, \omega_I)$ is the operator characteristic function of the heterodyne POVM.

We finally comment on the photon detectors used in quantum optics. On-off detectors (also called threshold detectors) distinguish zero and non-zero photons. Photon-number-resolving detectors distinguish up to a few photons (with current technology). Both detectors should distinguish photons with sufficiently high probability; i.e., the efficiency of the detector should be high. These detectors respond to the input photon number very non-linearly and thus are non-Gaussian operations. An example of the high-efficiency on-off detector used in quantum optics experiments is the superconducting nanowire single-photon detector (SNSPD), and an example of the photon-number-resolving detector is the transition edge sensor (TES). Both detectors utilize the transition event of the superconductor when absorbing photons. For details, see e.g. [37, 38] and the references therein.

References

1. Holevo AS (2011) Probabilistic and statistical aspects of quantum theory, Vol 1. Springer Science & Business Media
2. Braunstein SL, van Loock P (2005) Quantum information with continuous variables. Rev Mod Phys 77:513–577. https://doi.org/10.1103/RevModPhys.77.513
3. Eisert J, Plenio MB (2003) Introduction to the basics of entanglement theory in continuous-variable systems. Int J Quantum Inf 1(04):479–506
4. Cahill KE, Glauber RJ (1969) Density operators and quasiprobability distributions. Phys Rev 177:1882–1902. https://doi.org/10.1103/PhysRev.177.1882
5. Cerf NJ, Leuchs G, Polzik ES (2007) Quantum information with continuous variables of atoms and light. World Scientific
6. Weedbrook C et al (2012) Gaussian quantum information. Rev Mod Phys 84:621–669. https://doi.org/10.1103/RevModPhys.84.621
7. Simon R, Mukunda N, Dutta B (1994) Quantum-noise matrix for multimode systems: U(n) invariance, squeezing, and normal forms. Phys Rev A 49:1567–1583. https://doi.org/10.1103/PhysRevA.49.1567
8. Caruso F et al (2008) Multi-mode bosonic Gaussian channels. New J Phys 10(8):083030. https://doi.org/10.1088/1367-2630/10/8/083030
9. Holevo AS, Werner RF (2001) Evaluating capacities of bosonic Gaussian channels. Phys Rev A 63:032312. https://doi.org/10.1103/PhysRevA.63.032312
10. Caruso F, Giovannetti V, Holevo AS (2006) One-mode bosonic Gaussian channels: a full weak-degradability classification. New J Phys 8(12):310–310. https://doi.org/10.1088/1367-2630/8/12/310
11. Holevo AS (2007) One-mode quantum Gaussian channels: structure and quantum capacity. Probl Inf Transm 43(1):1–11. issn: 1608-3253. https://doi.org/10.1134/S0032946007010012
12. Solomon Ivan J, Sabapathy KK, Simon R (2011) Operator-sum representation for bosonic Gaussian channels. Phys Rev A 84:042311. https://doi.org/10.1103/PhysRevA.84.042311
13. Giovannetti V, Holevo AS, García-Patrón R (2015) A solution of gaussian optimizer conjecture for quantum channels. Commun Math Phys 334(3):1553–1571. issn: 1432-0916. https://doi.org/10.1007/s00220-014-2150-6
14. Caruso F, Giovannetti V (2006) Degradability of Bosonic Gaussian channels. Phys Rev A 74:062307. https://doi.org/10.1103/PhysRevA.74.062307
15. Caruso F et al (2011) Optimal unitary dilation for bosonic Gaussian channels. Phys Rev A 84:022306. https://doi.org/10.1103/PhysRevA.84.022306
16. Wolf MM, Pérez-García D, Giedke G (2007) Quantum Capacities of Bosonic channels. Phys Rev Lett 98:130501. https://doi.org/10.1103/PhysRevLett.98.130501
17. Pirandola S et al (2009) Direct and reverse secret-key capacities of a quantum channel. Phys Rev Lett 102:050503. https://doi.org/10.1103/PhysRevLett.102.050503
18. García-Patrón R et al (2012) Majorization theory approach to the gaussian channel minimum entropy conjecture. Phys Rev Lett 108:110505. https://doi.org/10.1103/PhysRevLett.108.110505
19. Wilde MM, Hayden P, Guha S (2012) Quantum trade-off coding for bosonic communication. Phys Rev A 86:062306. https://doi.org/10.1103/PhysRevA.86.062306
20. Mari A, Giovannetti V, Holevo AS (2014) Quantum state majorization at the output of bosonic Gaussian channels. Nat Commun 5(1):3826. issn: 2041-1723. https://doi.org/10.1038/ncomms4826
21. Giovannetti V et al (2014) Ultimate classical communication rates of quantum optical channels. Nat Photonics 8(10):796–800. issn: 1749- 4893. https://doi.org/10.1038/nphoton.2014.216
22. Takeoka M, Guha S, Wilde MM (2014) Fundamental rate-loss tradeoff for optical quantum key distribution. Nat Commun 5(1):5235. issn: 2041-1723. https://doi.org/10.1038/ncomms6235
23. De Palma G, Trevisan D, Giovannetti V (2016) Passive states optimize the output of Bosonic Gaussian quantum channels. IEEE Trans Inf Theory 62(5):2895–2906. https://doi.org/10.1109/TIT.2016.2547426

24. Pirandola S et al (2017) Fundamental limits of repeaterless quantum communications. Nat Commun 8(1):15043. issn: 2041-1723. https://doi.org/10.1038/ncomms15043
25. De Palma G, Trevisan D, Giovannetti V (2017) Gaussian states minimize the output entropy of the one-mode quantum attenuator. IEEE Trans Inf Theory 63(1):728–737. https://doi.org/10.1109/TIT.2016.2621748
26. Qi H, Wilde MM (2017) Capacities of quantum amplifier channels. Phys Rev A 95:012339. https://doi.org/10.1103/PhysRevA.95.012339
27. Wilde MM, Qi H (2018) Energy-constrained private and quantum capacities of quantum channels. IEEE Trans Inf Theory 64(12):7802–7827. https://doi.org/10.1109/TIT.2018.2854766
28. Sharma K et al (2018) Bounding the energy-constrained quantum and private capacities of phase-insensitive bosonic Gaussian channels. New J Phys 20(6):063025
29. Rosati M, Mari A, Giovannetti V (2018) Narrow bounds for the quantum capacity of thermal attenuators. Nat Commun 9(1):4339. issn: 2041-1723. https://doi.org/10.1038/s41467-018-06848-0
30. Holevo AS (2019) Quantum systems, channels, information: a mathematical introduction. De Gruyter. isbn: 9783110642490. https://doi.org/10.1515/9783110642490
31. Holevo AS (2021) Structure of a general quantum Gaussian observable. Proc Steklov Inst Math 313(1):7–77. issn: 1531-8605. https://doi.org/10.1134/S0081543821020085
32. Caves CM, Drummond PD (1994) Quantum limits on bosonic communication rates. Rev Mod Phys 66:481–537. https://doi.org/10.1103/RevModPhys.66.481
33. Mandel L, Wolf E (1995) Optical coherence and quantum optics. Cambridge University Press, Cambridge
34. Kok P, Lovett BW (2010) Introduction to optical quantum information processing. Cambridge University Press, Cambridge
35. Buchholz D et al (2016) The universal C*-Algebra of the electromagnetic field. Lett Math Phys 106(2):269–285. issn: 1573-0530. https://doi.org/10.1007/s11005-015-0801-y
36. Reck M et al (1994) Experimental realization of any discrete unitary operator. Phys Rev Lett 73:58–61. https://doi.org/10.1103/PhysRevLett.73.58
37. Irwin KD, Hilton GC (2005) Transition-edge sensors. In: Cryogenic particle detection, pp 63–150
38. Natarajan CM, Tanner MG, Hadfield RH (2012) Superconducting nanowire single-photon detectors: physics and applications. Supercond Sci Technol 25(6):063001

Chapter 4
Quantum Key Distribution with Continuous-Variable Systems

Abstract Continuous-variable quantum key distribution has affinity to the optical telecommunication, and thus has advantages in the real-world implementation. However, for continuous-variable quantum key distribution protocols that are compatible with digital signal processing, their finite-size security against general attacks has been a long-standing open problem. In this chapter, we solve the problem by developing a fidelity estimation that converges quickly in a finite number of trials and using it as a test for the information leakage.

Keywords Continuous-variable quantum key distribution · Finite-size security proof · Phase errors

4.1 Introduction for This Chapter

Quantum key distribution (QKD) aims at generating a secret key shared between two remote legitimate parties with information-theoretic security. QKD combined with the one-time pad [5] provides secure communication against an adversary with arbitrary computational power and hardware technology. Since the first proposal in 1984 [1], various QKD protocols have been proposed with many varieties of encoding and decoding schemes. These protocols are typically classified into two categories depending on the detection methods. One of them is called discrete-variable QKD, which uses photon detectors and includes earlier protocols such as BB84 [1] and B92 [6] protocols. The other is called continuous-variable QKD, which uses homodyne and heterodyne measurements with photodetectors [2, 7–12]. See Refs. [13, 14] for recent comprehensive reviews on the topic.

Although discrete-variable QKD is more mature and achieves longer-distant key distribution if photon detectors with low dark count rates are available, continuous-variable QKD has its own distinct advantages for a short distance. It can be implemented with components common to coherent optical communication technology

T. Matsuura, *Digital Quantum Information Processing with Continuous-Variable Systems*, Springer Theses, https://doi.org/10.1007/978-981-19-8288-0_4

and is expected to be cost-effective. Excellent spectral filtering capability inherent in homodyne/heterodyne measurements suppresses crosstalk in wavelength division multiplexing (WDM) channels. This allows multiplexing of hundreds of QKD channels into a single optical fiber [15] as well as co-propagation with classical data channels [16–23], which makes integration into an existing communication network easier.

One major obstacle in putting continuous-variable QKD to practical use is the gap between the employed continuous variables and mandatory digital signal processing. The continuous-variable QKD protocols using coherent states are divided into two branches depending on whether the modulation of the encoder is also continuous or discrete. The continuous modulation protocols usually adopts Gaussian modulation, in which the sender chooses the complex amplitude of a coherent-state pulse according to a Gaussian distribution [7, 8, 10–12] (see Ref. [24, 25] for a review). This allows powerful theoretical tools of the de Finetti theorem [26, 27], and complete security proofs for a finite-size key and against general attacks have been given [28–30]. In order to implement Gaussian protocols with a digital random-number generator and digital signal processing, it is necessary to approximate the continuous distribution with a constellation composed of a finite number of complex amplitudes. It turns out that an overwhelming number of coherent states is needed to directly approximate the Gaussian ensemble in order for the security condition to be satisfied [31, 32]. If we try to mitigate the required number, additional assumptions are needed, which makes it difficult to apply it in the finite-size regime [33]. The other branch gives priority to the simplicity of the modulation and uses a very small (usually two to four) number of amplitudes [34–37]. As for the security analysis, the status is more or less similar to the Gaussian constellation case, and current security proofs are either in the asymptotic regime against collective attacks [38–43] or in the finite-size regime but against more restrictive attacks [44, 45]. Hence, regardless of approaches, a complete security proof of continuous-variable QKD in the finite-size regime against general attacks has been a milestone yet to be achieved.

Here, we achieve the above milestone by proposing a binary phase-modulated continuous-variable QKD protocol with a composable security proof in the finite-size regime against general attacks. The key ingredient is an estimation method of the fidelity to the coherent states we develop here using the heterodyne measurement and the classical post-processing, which is suited for the analysis in the finite-size regime. Once the estimation method is developed, the security proof is accomplished by a reduction to the entanglement distillation, which is the established technique in the discrete-variable QKD.

The so-obtained security proof is further refined based on the reverse reconciliation, aiming at improving the performance of the protocol at the cost of more complexity in the security analysis. With a refined proof and no additional experimental requirement, we obtain a significant improvement in the key gain rate against loss. In fact, it achieves near-optimal scaling against transmission distance in the limit of infinite code length. These results accelerate the real-world implementation of reliable and provably-secure continuous-variable QKDs.

This chapter is organized as follows. In Sect. 4.2, the notations and preliminaries used in this chapter are summarized. In Sect. 4.3, the definition of the composable security of the QKD is stated, and an approach for the security proof is reviewed. Sections 4.4 and 4.5 are the results of this thesis. In Sect. 4.4, the composable security for a binary modulation continuous-variable QKD protocol is proved against general attacks in the finite-size regime. Our security proof is adapted to digital information processing and thus allows the use of the binned homodyne and heterodyne measurements (i.e., measurements with finite resolutions). The security proof is further refined in Sect. 4.5 based on the reverse reconciliation, and, as a result, the asymptotic key rate of the protocol achieves almost optimal scaling against transmission distance for the pure-loss channel (Sect. 4.4 is based on the publication [46].).

4.2 Notations and Preliminaries

In what follows, the base of the logarithm is taken to be 2.

4.2.1 Finite Field $\mathbb{F}_2$

Classical information processing is done on binary numbers. Binary numbers 0, 1 forms a finite field $\mathbb{F}_2$ with the addition $\oplus$ and the multiplication $\cdot$ that are defined as follows:

$$0 \oplus 0 = 1 \oplus 1 = 0, \quad 0 \oplus 1 = 1 \oplus 0 = 1, \tag{4.1}$$

$$0 \cdot 0 = 0 \cdot 1 = 1 \cdot 0 = 0, \quad 1 \cdot 1 = 1. \tag{4.2}$$

In what follows, arithmetics of binary numbers (i.e., bits) follow the above.

An N-bit row vector $\boldsymbol{u}$ is an element of $\mathbb{F}_2^N$. The inner product $\boldsymbol{u}\boldsymbol{v}^\top$ of two N-bit row vectors $\boldsymbol{u}, \boldsymbol{v} \in \mathbb{F}_2^N$ is defined as

$$\boldsymbol{u}\boldsymbol{v}^\top = \bigoplus_{i=1}^{N} u_i \cdot v_i. \tag{4.3}$$

For a $N \times N$ binary matrix C, $\boldsymbol{u}C$ is also an N-bit row vector. The Hamming weight $\mathrm{wt}(\boldsymbol{u})$ of $\boldsymbol{u} \in \mathbb{F}_2^N$ is defined as

$$\mathrm{wt}(\boldsymbol{u}) = \left|\{i \in \{1, \ldots, N\} : u_i = 1\}\right|. \tag{4.4}$$

4.2.2 Classical Linear Information Processing as a Quantum Operation

Let $\boldsymbol{u}$ be an N-bit row vector and C be a $N \times N$ non-singular binary matrix. Let $|\boldsymbol{u}\rangle := |u_1\rangle \otimes \cdots \otimes |u_N\rangle$ be the pure quantum state on the N qubit system, where each qubit is in one of the Z-basis states $\{|0\rangle, |1\rangle\}$. If $|\boldsymbol{u}\rangle$ is measured on the Z basis, the binary string $\boldsymbol{u}$ is obtained with unit probability. The unitary operation that corresponds to the action of C on the Z basis of $|\boldsymbol{u}\rangle$ is defined as

$$U(C) = \sum_{z \in \{0,1\}^N} |zC\rangle\langle z| = \sum_{x \in \{0,1\}^N} |\boldsymbol{x}(C^\top)^{-1}{}_X\rangle\langle \boldsymbol{x}_X| . \tag{4.5}$$

Then, if $U(C)\,|\boldsymbol{u}\rangle$ is measured on the Z basis, $\boldsymbol{u}C$ is obtained with unit probability. One can derive the second equality of Eq. (4.5) by the following Fourier transform between the Z- and X-basis:

$$|z\rangle = \frac{1}{\sqrt{2^N}} \sum_{x \in \{0,1\}^N} (-1)^{zx^\top} |\boldsymbol{x}_X\rangle . \tag{4.6}$$

An example of $U(C)$ is the Controlled-NOT (CNOT) operation with C being

$$C = \begin{pmatrix} 1 & 1 \\ 0 & 1 \end{pmatrix} = C^{-1}. \tag{4.7}$$

Let $|\boldsymbol{u}\rangle = |u_1\rangle |u_2\rangle$ and $|\boldsymbol{v}_X\rangle = |v_{1X}\rangle |v_{2X}\rangle$ be pure quantum states with the binary strings $\boldsymbol{u}$ and $\boldsymbol{v}$ encoded on the Z and X basis, respectively. Then, the above CNOT gate acts as

$$\text{CNOT}\,|\boldsymbol{u}\rangle = |u_1\rangle\,|u_1 \oplus u_2\rangle , \tag{4.8}$$

where the first qubit is the control qubit and the second qubit is the target qubit. Using Eq. (4.5), we also have

$$\text{CNOT}\,|\boldsymbol{v}_X\rangle = |(v_1 \oplus v_2)_X\rangle\,|v_{2X}\rangle . \tag{4.9}$$

Let $\boldsymbol{v}$ be another N-bit row vector. Then, $\boldsymbol{u} \oplus \boldsymbol{v}$ is also a N-bit row vector. The unitary that corresponds to the addition of $\boldsymbol{v}$ on the Z basis is given by

$$U^{\oplus}(\boldsymbol{v}) = \sum_{z \in \{0,1\}^N} |z \oplus \boldsymbol{v}\rangle\langle z| , \tag{4.10}$$

If $U^{\oplus}(\boldsymbol{v})\,|\boldsymbol{u}\rangle$ is measured on the Z basis, $\boldsymbol{u} \oplus \boldsymbol{v}$ is obtained with unit probability. One can also check from (4.6) that this unitary acts as an identity on the X basis (up to the irrelevant global phase). In the same way, $U_X^{\oplus}(\boldsymbol{v}')$ is the unitary that corresponds to the addition of $\boldsymbol{v}'$ on the X basis. The simple examples of $U^{\oplus}(\boldsymbol{v})$ and $U_X^{\oplus}(\boldsymbol{v}')$

when $N = 1$ are the Pauli-X σ^X and the Pauli-Z σ^Z operators, which can be written respectively as

$$\sigma^X = |0\rangle\langle 1| + |1\rangle\langle 0|, \qquad \sigma^Z = |0_X\rangle\langle 1_X| + |1_X\rangle\langle 0_X|. \tag{4.11}$$

To sum up, one can rewrite classical linear information processing such as multiplying C and adding $\boldsymbol{v}$ on the binary sequence $\boldsymbol{u}$ as the unitary actions such as acting $U(C)$ and $U^{\oplus}(\boldsymbol{v})$ on the quantum state $|\boldsymbol{u}\rangle$ (followed by the Z-basis measurement).

4.2.3 Definitions and Properties of the Entropic Quantities

The definitions and properties of the entropic quantities used in this chapter are listed. $\mathbb{E}_P[\check{X}]$ denotes the expectation value of $\check{X}$, where $\check{X} : \Omega \to \mathbb{R}$ denotes the random variable for the probability space (Ω, Σ, P). Throughout the chapter, random variables are denoted with the symbol $\check{}$. In this section, only the discrete random variable is treated, and thus P_X denotes the probability mass function of $\check{X}$ and satisfies $P_X(x) = P(\check{X} = x) = P(\check{X}^{-1}(x))$.

Definition 4.2.1 (Entropy function) Let $\check{X}$ be a random variable with possible values in $\mathcal{X} := \{x_1, \ldots, x_n\}$, and P_X be the probability mass function of $\check{X}$. Then, the entropy function $H(\check{X})_{P_X}$ of P_X is defined as

$$H(\check{X})_{P_X} := \mathbb{E}_{P_X}[-\log P_X] = \sum_{i=1}^{n} -P_X(x_i) \log P_X(x_i). \tag{4.12}$$

The entropy function is non-negative. The entropy function is concave; i.e., for probability mass functions P_1, P_2 of $\check{X}$ and a number $p \in [0, 1]$, the following inequality holds:

$$H(\check{X})_{pP_1+(1-p)P_2} \geq pH(\check{X})_{P_1} + (1-p)H(\check{X})_{P_2}. \tag{4.13}$$

Definition 4.2.2 (*Binary entropy function*) The binary entropy function is defined as

$$h(p) := -p \log p - (1-p) \log(1-p). \tag{4.14}$$

The binary entropy function $h(p)$ is a special case of the entropy function; i.e., for a binary random variable $\check{X}$ ($\mathcal{X} = \{0, 1\}$) with $P_X(0) = p$, we have $H(\check{X})_{P_X} = h(p)$. It can be shown that $0 \leq h(p) \leq 1$. The following lemma with respect to the binary entropy function is proved here for later use.

Lemma 4.2.3 *(Upper-bound on the number of possible patterns) Let* $\boldsymbol{e} \in \{0, 1\}^n$ *be the n-bit sequence and* $\mathrm{wt}(\boldsymbol{e})$ *be the Hamming weight of* $\boldsymbol{e}$*. For* $0 \leq m \leq n$*, let* $\mathcal{T}_m$

be the set of n-bit sequences defined as $\mathcal{T}_m := \{e \in \{0,1\}^n : \mathrm{wt}(e) \le m\}$. *Then, the following inequality holds:*

$$|\mathcal{T}_m| \le 2^{nh(p)}, \quad p := \min\{m/n,\ 1/2\}, \tag{4.15}$$

where $|\mathcal{T}_m|$ *denotes the cardinality of the set* $\mathcal{T}_m$.

Proof It is trivial when $p = 1/2$ (i.e., $m/n \ge 1/2$). Therefore, we prove the case $p = m/n < 1/2$. Let $\Pr(e)$ be defined as $\Pr(e) := p^{\mathrm{wt}(e)}(1-p)^{n-\mathrm{wt}(e)}$. Since, for $p < 1/2$, $\Pr(e)$ monotonically decreases with respect to $\mathrm{wt}(e)$, $\Pr(e) \ge p^m(1-p)^{n-m} = 2^{-nh(m/n)}$ holds. From this, $1 \ge \sum_{e\in\mathcal{T}_m} \Pr(e) \ge |\mathcal{T}_m| 2^{-nh(m/n)}$ follows, which proves the statement of the proposition. □

Definition 4.2.4 (*Kullback-Leibler divergence (relative entropy function)*) Let $\check{X}$ be a random variable that takes value in the set $\mathcal{X} := \{x_1, \ldots, x_n\}$. Let P_X and Q_X be probability mass functions of $\check{X}$. Then the Kullback-Leibler divergence $D(P_X \| Q_X)$ is defined as

$$D(P_X\|Q_X) := \sum_{i=1}^{n} P_X(x_i) \log \frac{P_X(x_i)}{Q_X(x_i)} = -\sum_{i=1}^{n} P_X(x_i) \log \frac{Q_X(x_i)}{P_X(x_i)}. \tag{4.16}$$

Lots of important properties and inequalities for entropic functions can be derived from the properties of the Kullback-Leibler divergence. Examples of such properties are:

- (Asymmetry) $D(P_X\|Q_X)$ is in general asymmetric; i.e. $D(P_X\|Q_X) = D(Q_X\|P_X)$ does not necessarily hold. This is obvious from the definition.
- (Non-negativity) $D(P_X\|Q_X)$ satisfies $D(P_X\|Q_X) \ge 0$. This can be proved by applying Jensen's inequality to the logarithmic function:

$$D(P_X\|Q_X) = -\sum_{i=1}^{n} P_X(x_i) \log \frac{Q_X(x_i)}{P_X(x_i)} \ge -\log\left(\sum_{i=1}^{n} P_X(x_i) \frac{Q_X(x_i)}{P_X(x_i)}\right) = 0. \tag{4.17}$$

- (Joint convexity) For probability mass functions P_1, P_2, Q_1, Q_2 and a number $p \in [0,1]$, the Kullback-Leibler divergence satisfies the following inequality:

$$D\left(pP_1 + (1-p)P_2 \| pQ_1 + (1-p)Q_2\right) \le pD(P_1\|Q_1) + (1-p)D(P_2\|Q_2). \tag{4.18}$$

  The proof of this property is given in e.g. [47].
- (Data processing) Let $\check{X}$ and $\check{Y}$ be random variables that take values in $\mathcal{X}$ and $\mathcal{Y}$, respectively, and let $(\Lambda(y \mid x))_{x\in\mathcal{X}, y\in\mathcal{Y}}$ be the transition matrix for the stochastic process $\check{X} \to \check{Y}$, that is, each matrix element $\Lambda(y \mid x)$ is non-negative and $\sum_{y\in\mathcal{Y}} \Lambda(y \mid x) = 1$. Let P_X, Q_X be probability mass functions of $\check{X}$. Then the following inequality holds (data processing inequality):

$$D(P_X \| Q_X) \geq D(P_Y \| Q_Y), \tag{4.19}$$

where $P_Y(y) := \sum_{x \in \mathcal{X}} \Lambda(y \mid x) P_X(x)$, $Q_Y(y) := \sum_{x \in \mathcal{X}} \Lambda(y \mid x) Q_X(x)$. The proof is given in e.g. [47]. The data processing inequality is a very important property in information theory.

Now we move on to the quantum entropic quantities, which are not directly used in this chapter but nevertheless important in the theory of security proof of quantum key distribution. Here, we assume that the quantum system we treat is finite-dimensional. First, we define a generalization of the fidelity to subnormalized states.

Definition 4.2.5 (*Generalized fidelity*) For subnormalized states ρ, σ (i.e., $\rho, \sigma \geq 0$ and $\mathrm{Tr}(\rho), \mathrm{Tr}(\rho) \leq 1$), the generalized fidelity $\tilde{F}(\rho, \sigma)$ is defined as

$$\tilde{F}(\rho, \sigma) := \left(\| \sqrt{\rho} \sqrt{\sigma} \|_1 + \sqrt{(1 - \mathrm{Tr}(\rho))(1 - \mathrm{Tr}(\sigma))} \right)^2 . \tag{4.20}$$

By definition, the generalized fidelity is reduced to the usual fidelity (Definition 2.2.6) when the states are normalized.

Definition 4.2.6 (*Conditional min- and max-entropies* [3, 48–51]) Let ρ_{AB} be a subnormalized state on $\mathcal{H}_A \otimes \mathcal{H}_B$ (i.e., $\mathrm{Tr}[\rho_{AB}] \leq 1$). Then, the min- and max-entropies of A conditioned on B of ρ_{AB} are given respectively by

$$H_{\min}(A|B)_\rho := \max_{\sigma \in \mathfrak{D}(\mathcal{H}_B)} \sup\{\lambda \in \mathbb{R} : \rho_{AB} \leq 2^{-\lambda} I_A \otimes \sigma_B\} \tag{4.21}$$

$$= -\log \min_{\sigma \geq 0} \{\mathrm{Tr}(\sigma) : \rho_{AB} \leq I_A \otimes \sigma_B\}, \tag{4.22}$$

$$H_{\max}(A|B)_\rho := \max_{\sigma \in \mathfrak{D}(\mathcal{H}_B)} \log \left(d_A \tilde{F}(\rho_{AB}, d_A^{-1} I_A \otimes \sigma_B) \right), \tag{4.23}$$

where d_A denotes the dimension of $\mathcal{H}_A$.

In order to introduce the smoothed version of the conditional min- and max- entropies, we introduce another distance measure for the subnormalized states.

Definition 4.2.7 (*Purified distance* [3]) For subnormalized states ρ, σ, the purified distance $P(\rho, \sigma)$ is defined as

$$P(\rho, \sigma) := \sqrt{1 - \tilde{F}(\rho, \sigma)}. \tag{4.24}$$

Definition 4.2.8 (*Smooth conditional min- and max-entropies* [3]) For a subnormalized state ρ_{AB} and $\varepsilon \geq 0$, the ε-smooth min- and max-entropies of A conditioned on B of ρ_{AB} are given respectively by

$$H^\varepsilon_{\min}(A|B)_\rho := \max_{\tilde{\rho} \in \mathcal{B}^\varepsilon(\rho_{AB})} H_{\min}(A|B)_{\tilde{\rho}}, \tag{4.25}$$

$$H^\varepsilon_{\max}(A|B)_\rho := \max_{\tilde{\rho} \in \mathcal{B}^\varepsilon(\rho_{AB})} H_{\max}(A|B)_{\tilde{\rho}}, \tag{4.26}$$

where the ε-ball $\mathcal{B}^{\varepsilon}(\rho_{AB})$ is defined as

$$\mathcal{B}^{\varepsilon}(\rho_{AB}) := \{\tau \in \mathfrak{T}(\mathcal{H}_{AB}) : \tau \geq 0, \mathrm{Tr}(\tau) \leq 1, P(\tau, \rho) \leq \varepsilon\}. \tag{4.27}$$

4.2.4 Concentration Inequalities

In this section, concentration inequalities used in this thesis are listed. The probability space (Ω, Σ, P) and the associated random variable $\check{X}$ we treat here can be continuous. The first one is the so-called Chernoff bound, which is a tail bound for the binomial distribution.

Theorem 4.2.9 (Tail bound for the binomial distribution [52–54]) *Let $\check{X}_1, \ldots, \check{X}_n$ be independent and identically distributed binary random variables with $P(\check{X}_i = 1) = p$, for $i = 1, \ldots, n$. Then, for any $\delta \in [0, 1-p]$, the following inequality holds:*

$$P\left(\frac{1}{n}\sum_{i=1}^{n}\check{X}_i > p + \delta\right) < 2^{-nD(p+\delta\|p)}, \tag{4.28}$$

where $D(p+\delta\|p)$ denotes

$$D(p+\delta\|p) := (p+\delta)\log\frac{p+\delta}{p} + (1-p-\delta)\log\frac{1-p-\delta}{1-p}. \tag{4.29}$$

The above is the bound of the probability for the upper tail. The bound for the lower tail can be obtained similarly by exchanging the values of the random variables as well as changing p to $1-p$. The following corollary of the above theorem is suitable for our applications.

Corollary 4.2.10 (Alternative form of the tail bound) *Let $\check{X}_1, \ldots, \check{X}_n$ be independent and identically distributed binary random variables with $P(\check{X}_i = 1) = p$, for $i = 1, \ldots, n$. Given $\epsilon \in (0, 1)$, we have*

$$P\left(\frac{1}{n}\sum_{i=1}^{n}\check{X}_i > p + \delta(\epsilon; n)\right) < \epsilon, \tag{4.30}$$

where $\delta(\epsilon; n)$ is defined to satisfy

$$\begin{cases} D(p+\delta(\epsilon; n)\|p) = -\frac{1}{n}\log\epsilon & (\epsilon > p^n) \\ \delta(\epsilon; n) = 1-p & (\epsilon \leq p^n) \end{cases}. \tag{4.31}$$

Proof Since $D(p+\delta\|p)$ increases monotonically from 0 to $-\log p$ when δ varies from 0 to 1, Theorem 4.2.9 can be applied for $\epsilon > p^n$ by the suitable choice of

$\delta(\epsilon; n)$. On the other hand, $P(\frac{1}{n}\sum_{i=1}^{n}\check{X}_i > 1) = 0 (< \epsilon)$ always holds. These prove the statement of the above Corollary. □

The Chernoff-type bound also holds for the hypergeometric distribution. The following is the statement.

Theorem 4.2.11 (Tail bound for the hypergeometric distribution [55]) *Let $X_1, \ldots, X_N$ be a binary sequence, and M be the number of elements with $X_i = 1$, i.e, $M := \sum_{i=1}^{N} X_i$. Let $\check{Y}_1, \ldots, \check{Y}_n$ be randomly sampled from $X_1, \ldots, X_N$ without replacement. Let $\check{m} := \sum_{i=1}^{n} \check{Y}_i$ be the number of ones in $\check{Y}_1, \ldots, \check{Y}_n$. Then, for any $\delta \in [0, M/N]$, the following inequality holds:*

$$P\left(\frac{\check{m}}{n} \le \frac{M}{N} - \delta\right) \le 2^{-nD\left(\frac{M}{N}-\delta \middle\| \frac{M}{N}\right)}, \tag{4.32}$$

where $D(\cdot\|\cdot)$ is defined in Eq. (4.29).

The above is the bound of the probability for the lower tail. The bound for the upper tail can be obtained similarly. The following corollary is useful for the bit error sampling.

Corollary 4.2.12 (Estimation by the simple random sampling without replacement) *Let $X_1, \ldots, X_N$ be a binary sequence with $M := \sum_{i=1}^{N} X_i$. Let $\check{Y}_1, \ldots, \check{Y}_n$ be randomly sampled from $X_1, \ldots, X_N$ without replacement, and define $\check{m} := \sum_{i=1}^{n} \check{Y}_i$. Then, for any $\epsilon \in (0, 1)$, the following inequality holds:*

$$P\left(\hat{M}_{N,n,\epsilon}(\check{m}) < M\right) \le \epsilon, \tag{4.33}$$

where the function $\hat{M}_{N,n,\epsilon}(m)$ is defined to satisfy

$$\frac{m}{n} \le \frac{\hat{M}_{N,n,\epsilon}(m)}{N} \le 1 \tag{4.34}$$

and for $m \in [0, n)$,

$$D\left(m/n \middle\| \hat{M}_{N,n,\epsilon}(m)/N\right) = -\frac{1}{n}\log \epsilon. \tag{4.35}$$

Proof Let $f(M)$ be a function of M satisfying $0 \le f(M)/n \le M/N$. Then, from Theorem 4.2.11, we have

$$P\left(\frac{\check{m}}{n} \le \frac{M}{N} - \left(\frac{M}{N} - \frac{f(M)}{n}\right)\right) \le 2^{-nD\left(\frac{f(M)}{n} \middle\| \frac{M}{N}\right)}. \tag{4.36}$$

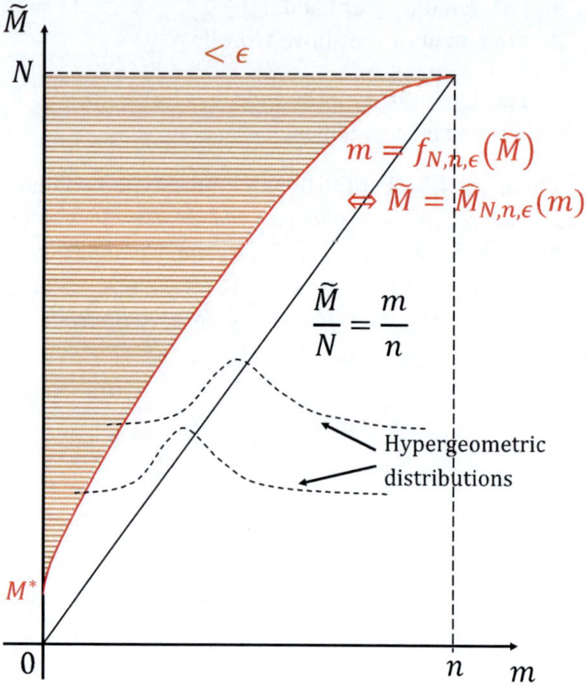

Fig. 4.1 The definition of the function $f_{N,n,\epsilon}(\tilde{M})$ or equivalently $\hat{M}_{N,n,\epsilon}(m)$ in Corollary 4.2.12. The real number M^* is the intercept of the function $\hat{M}_{N,n,\epsilon}(m)$, i.e., $M^* = \hat{M}_{N,n,\epsilon}(0)$. Given N, n, and ϵ, the failure event for each M, i.e., $\hat{M}_{N,n,\epsilon}(\check{m}) \leq M$, occurs when $\check{m} \leq f_{N,n,\epsilon}(M)$. In order to bound the probability of such events, we can use the tail bound for the hypergeometric distribution in Theorem 4.2.11

We set the function $f(M)$ to the restriction of the function $f_{N,n,\epsilon}(\tilde{M})$ of the real number $\tilde{M}$ that satisfies

$$D\left(f_{N,n,\epsilon}(\tilde{M})/n \| \tilde{M}/N\right) = -\frac{1}{n}\log\epsilon, \tag{4.37}$$

for $((1 - \sqrt[n]{\epsilon})N =) M^* \leq \tilde{M} < N$ (see Fig. 4.1). (The case $\tilde{M} = N$ is singular, so we take the limit of the case $\tilde{M} = N - \delta$, $\delta \to 0$, which leads to $f_{N,n,\epsilon}(N - \delta) \to n$.) The function $f_{N,n,\epsilon}(\tilde{M})$ increases monotonically with increasing $\tilde{M}$ in $[M^*, N)$, and its range lies in $[0, n)$. Thus, from Eq. (4.36), we have

$$P\left(f^{-1}_{N,n,\epsilon}(\check{m}) \leq M\right) \leq \epsilon \tag{4.38}$$

for any $\check{m} \in [0, n)$. We define $\hat{M}_{N,n,\epsilon}(m) := f^{-1}_{N,n,\epsilon}(m)$ for $m \in [0, n)$, which satisfies Eqs. (4.34) and (4.35) by construction. To incorporate the case $\check{m} = n$, we use the following weaker version of Eq. (4.38) that trivially follows from Eq. (4.38) for $\check{m} \in [0, n)$:

$$P\left(\hat{M}_{N,n,\epsilon}(\check{m}) < M\right) \leq \epsilon. \tag{4.39}$$

Define $\hat{M}_{N,n,\epsilon}(n) = N$, which is the unique choice to satisfy Eq. (4.34) for $m = n$. Then, Eq. (4.39) holds also for $\check{m} = n$. This proves Eq. (4.33) with $\hat{M}_{N,n,\epsilon}$ satisfying Eqs. (4.34) and (4.35) by construction in Eq. (4.38) and definition of $\hat{M}_{N,n,\epsilon}(n) = N$. □

The Chernoff-type bound is tight in many cases, but its applicability is restricted. The following inequality is looser than the Chernoff-type bound but can be applied to more general distributions.

Theorem 4.2.13 (Hoeffding's inequality [53, 54]) *Let $\check{X}_1, \ldots, \check{X}_n$ be independent random variables with each $\check{X}_i$ bounded by the intervals $[a_i, b_i]$ almost surely. Then, for $t \geq 0$, we have*

$$P\left(\sum_{i=1}^{n}(\check{X}_i - \mathbb{E}[\check{X}_i]) \geq t\right) \leq \exp\left(-\frac{2t^2}{\sum_{i=1}^{n}(b_i - a_i)^2}\right). \tag{4.40}$$

Note that, for two random variables $\check{X}_1$ and $\check{X}_2$, the condition that $\check{X}_1 \geq \check{X}_2$ holds almost surely means that $P(\check{X}_1 \geq \check{X}_2) = 1$. Equivalently, $\check{X}_1 \geq \check{X}_2$ holds almost surely if and only if $\int_A (\check{X}_1 - \check{X}_2) dP \geq 0$ for any $A \in \Sigma$. Lots of refinement of the Hoeffding's inequality (e.g., by using the information of higher moments of the probability distributions) have been studied [53, 54].

In the following, we explain Azuma's inequality, which has a wide range of applications since, contrary to the previous bounds, the random variable can be neither independent nor identically distributed. First, we introduce the following notion.

Definition 4.2.14 (*$\mathcal{F}$-measurable*) Let (Ω, Σ, P) be a probability space. Let $\mathcal{F}$ be a σ-subalgebra of Σ. The random variable $\check{X}$ is called $\mathcal{F}$-measurable if, for any Borel set $A \in \mathcal{B}(\mathbb{R})$ (where $\mathcal{B}(\mathbb{R})$ denotes the Borel σ-algebra of $\mathbb{R}$), it satisfies

$$\check{X}^{-1}(A) := \{y \in \Omega \mid \check{X}(y) \in A\} \in \mathcal{F}. \tag{4.41}$$

In the case of the discrete probability space, the following condition is equivalent to the above. For $x \in \Omega$, let $\mathcal{F}(x)$ be the smallest set in $\mathcal{F}$ containing x i.e.,

$$\mathcal{F}(x) := \bigcap_{\substack{S \in \mathcal{F} \\ x \in S}} S. \tag{4.42}$$

Then, the random variable $\check{X}$ is $\mathcal{F}$-measurable if $\check{X}(x) = \check{X}(y)$ for any $x \in \Omega$ and $y \in \mathcal{F}(x)$. Note that, if $\check{X}$ is $\mathcal{G}$-measurable for $\mathcal{G} \subset \mathcal{F}$, then it is trivially $\mathcal{F}$-measurable. Note also that a constant random variable is measurable with respect to the trivial σ-algebra $\{\varnothing, \Omega\}$.

Next we introduce the conditional expectation.

Definition 4.2.15 (*Conditional expectation*) Let (Ω, Σ, P) and $\mathcal{F}$ be the same as above. Let $\check{X} : \Omega \to \mathbb{R}$ be a random variable with a finite expectation. Then, the conditional expectation $\mathbb{E}[\check{X} \mid \mathcal{F}] : \Omega \to \mathbb{R}$ is defined as an $\mathcal{F}$-measurable function satisfying

$$\int_F \mathbb{E}[\check{X} \mid \mathcal{F}] dP = \int_F \check{X}\, dP, \qquad \forall F \in \mathcal{F}. \tag{4.43}$$

The definition of the conditional expectation is more tractable in the case of the discrete probability space; $\mathbb{E}[\check{X} \mid \mathcal{F}] : \Omega \to \mathbb{R}$ is given by

$$\mathbb{E}[\check{X} \mid \mathcal{F}](x) = \frac{1}{\sum_{y \in \mathcal{F}(x)} P(y)} \sum_{y \in \mathcal{F}(x)} \check{X}(y) P(y). \tag{4.44}$$

This can be regarded as a derivative form of Eq. (4.43). In fact, the conditional expectation can be given by the Radon-Nikodym derivative of probability measures. The conditional expectation has several important properties. Here we list some of them that are relevant to this thesis.

- (Linearity) For two random variables $\check{X}_1, \check{X}_2$, a σ-algebra $\mathcal{F}$, and $a \in \mathbb{R}$, the following holds:

$$\mathbb{E}[\check{X}_1 + a\check{X}_2 \mid \mathcal{F}] = \mathbb{E}[\check{X}_1 \mid \mathcal{F}] + a\, \mathbb{E}[\check{X}_2 \mid \mathcal{F}]. \tag{4.45}$$

- (Monotonicity) For $\check{X}_1 \le \check{X}_2$ (almost surely), we have

$$\mathbb{E}[\check{X}_1 \mid \mathcal{F}] \le \mathbb{E}[\check{X}_2 \mid \mathcal{F}]. \tag{4.46}$$

- (Stability) For an $\mathcal{F}$-measurable random variable $\check{X}$, we have

$$\mathbb{E}[\check{X} \mid \mathcal{F}] = \check{X}. \tag{4.47}$$

- (Tower property) For σ-algebras $\mathcal{G} \subset \mathcal{F}$, the following holds:

$$\mathbb{E}\left[\mathbb{E}[\check{X} \mid \mathcal{F}] \;\middle|\; \mathcal{G} \right] = \mathbb{E}[\check{X} \mid \mathcal{G}]. \tag{4.48}$$

  Note that $\mathbb{E}\left[\mathbb{E}[\check{X} \mid \mathcal{G}] \;\middle|\; \mathcal{F} \right] = \mathbb{E}[\check{X} \mid \mathcal{G}]$ also holds from the stability.

Now we introduce the following.

Definition 4.2.16 (*Filtration*) A sequence of σ-algebras $\mathcal{F}_0, \mathcal{F}_1, \ldots, \mathcal{F}_n$ is called the filtration if

$$\{\varnothing, \Omega\} = \mathcal{F}_0 \subseteq \mathcal{F}_1 \subseteq \cdots \subseteq \mathcal{F}_n. \tag{4.49}$$

We finally introduce the following notion.

Definition 4.2.17 (*Martingale*) Let $\{\varnothing, \Omega\} = \mathcal{F}_0 \subseteq \mathcal{F}_1 \subseteq \cdots \subseteq \mathcal{F}_n$ be a filtration. The sequence of random variables $(\check{X}_1, \ldots, \check{X}_n)$ is called a martingale with respect to this filtration if $\check{X}_i$ is $\mathcal{F}_i$-measurable, and

$$\mathbb{E}[\check{X}_i \mid \mathcal{F}_{i-1}] = \check{X}_{i-1} \tag{4.50}$$

holds (almost surely) for $1 \leq i \leq n$. Here, $\check{X}_0$ is a constant function satisfying $\check{X}_0 = \mathbb{E}[\check{X}_1 \mid \mathcal{F}_0] = \mathbb{E}[\check{X}_1]$.

Equation (4.50) is an equality as functions of Ω. An example of the $\mathcal{F}_i$-measurable sequence in the case of the discrete probability space is as follows. Let $\mathcal{Y}$ be a finite set. Let $\Omega = \mathcal{Y}^{\times n}$ be an n-ary Cartesian product and $\mathcal{F}_i = \{\mathcal{X} \times \mathcal{Y}^{\times(n-i)} \mid \mathcal{X} \in \mathcal{P}(\mathcal{Y}^{\times i})\}$ be a σ-subalgebra, where $\mathcal{P}(\mathcal{Y}^{\times i})$ is the power set of $\mathcal{Y}^{\times i}$. Given the sequence of outcomes $(y_1, \ldots, y_n) \in \Omega$, the random variable $\check{X}_i$ should depend only on $(y_1, \ldots, y_i)$ to be $\mathcal{F}_i$-measurable; that is, $\check{X}_i$ is fixed when given an element of $\mathcal{F}_i$. On the other hand, $\check{X}_i$ is in general still random when given an element of $\mathcal{F}_{i-1}$. Note that, by definition, $\check{X}_i$ is $\mathcal{F}_k$-measurable for $k \geq i$ if it is $\mathcal{F}_i$-measurable.

Now we can state Azuma's inequality as follows.

Theorem 4.2.18 (Azuma's inequality [56–58]) *Let $(\check{X}_1, \ldots, \check{X}_n)$ be a martingale sequence with respect to the filtration $\{\varnothing, \Omega\} = \mathcal{F}_0 \subseteq \mathcal{F}_1 \subseteq \cdots \subseteq \mathcal{F}_n$. Suppose there are predictable processes $(a_1, \ldots, a_n)$ and $(b_1, \ldots, b_n)$ with respect to $\mathcal{F}_0 \subseteq \mathcal{F}_1 \subseteq \cdots \subseteq \mathcal{F}_n$ (i.e., a_i and b_i are $\mathcal{F}_{i-1}$-measurable for $1 \leq i \leq n$), such that*

$$a_i \leq \check{X}_i - \check{X}_{i-1} \leq b_i \tag{4.51}$$

holds (almost surely). Then, for $t \geq 0$, we have

$$P(\check{X}_n - \check{X}_0 \geq t) \leq \exp\left(-\frac{2t^2}{\sum_{i=1}^n (b_i - a_i)^2}\right). \tag{4.52}$$

The following corollary combines Azuma's inequality with the technique called the Doob decomposition.

Corollary 4.2.19 (Azuma's inequality combined with the Doob decomposition) *Let $\{\varnothing, \Omega\} = \mathcal{F}_0 \subseteq \mathcal{F}_1 \subseteq \cdots \subseteq \mathcal{F}_n$ be a filtration. Let $(a_1, \ldots, a_n)$ and $(b_1, \ldots, b_n)$ be predictable processes with respect to this filtration. Let $(\check{X}_1, \ldots, \check{X}_n)$ be the sequence of random variables such that for $1 \leq i \leq n$, $\check{X}_i$ is $\mathcal{F}_i$-measurable and satisfies*

$$a_i \leq \check{X}_i \leq b_i, \qquad \textit{(almost surely)}. \tag{4.53}$$

Then, for $t \geq 0$, we have

$$P\left(\sum_{i=1}^n (\check{X}_i - \mathbb{E}[\check{X}_i \mid \mathcal{F}_{i-1}]) \geq t\right) \leq \exp\left(-\frac{2t^2}{\sum_{i=1}^n (b_i - a_i)^2}\right). \tag{4.54}$$

Proof For $1 \leq i \leq n$, let $\check{Y}_i$ be defined as

$$\check{Y}_i := \sum_{k=1}^{i} (\check{X}_k - \mathbb{E}[\check{X}_k \mid \mathcal{F}_{k-1}]). \tag{4.55}$$

Then, the sequence $(\check{Y}_1, \ldots, \check{Y}_n)$ is a martingale with respect to the filtration $\mathcal{F}_0 \subseteq \mathcal{F}_1 \subseteq \cdots \subseteq \mathcal{F}_n$, since $\check{Y}_i$ is $\mathcal{F}_i$-measurable and satisfies

$$\mathbb{E}[\check{Y}_i \mid \mathcal{F}_{i-1}] = \sum_{k=1}^{i} \left(\mathbb{E}[\check{X}_k \mid \mathcal{F}_{i-1}] - \mathbb{E}\Big[\mathbb{E}[\check{X}_k \mid \mathcal{F}_{k-1}] \Bigm| \mathcal{F}_{i-1} \Big] \right) \tag{4.56}$$

$$= \sum_{k=1}^{i-1} (\check{X}_k - \mathbb{E}[\check{X}_k \mid \mathcal{F}_{k-1}]) \tag{4.57}$$

$$= \check{Y}_{i-1}. \tag{4.58}$$

The second equality follows from the fact that $\check{X}_k$ is $\mathcal{F}_{i-1}$ measurable for $1 \leq k \leq i-1$ and the stability Eq. (4.47) of the conditional expectation. Note that $Y_0 := 0$. Furthermore, $(\check{Y}_1, \ldots, \check{Y}_n)$ satisfies

$$a_i - \mathbb{E}[\check{X}_i \mid \mathcal{F}_{i-1}] \leq \check{Y}_i - \check{Y}_{i-1} \leq b_i - \mathbb{E}[\check{X}_i \mid \mathcal{F}_{i-1}], \tag{4.59}$$

(almost surely) for $1 \leq i \leq n$. Since $a_i - \mathbb{E}[\check{X}_i \mid \mathcal{F}_{i-1}]$ and $b_i - \mathbb{E}[\check{X}_i \mid \mathcal{F}_{i-1}]$ are $\mathcal{F}_{i-1}$-measurable, we can apply Theorem 4.2.18 to the sequence $(\check{Y}_1, \ldots, \check{Y}_n)$ and prove the statement. □

At the cost of versatility, Azuma's inequality is not tight for most practical applications. The recent refinement of Azuma's inequality that is suitable for QKD applications is given in Ref. [59].

4.3 The Basics of the QKD

4.3.1 The Goal of the QKD

The goal of the QKD is to distribute the secret key between the distant two parties connected by a quantum channel and an authenticated public classical channel while ensuring that the key is secure against any eavesdropping attacks allowed by the law of quantum mechanics. The "secret key" here means a shared sequence of random numbers, and the "security" needs to meet the following two conditions; (1) two parties share the coincident keys (correctness condition), and (2) the key is secret against all except two parties (secrecy condition). With these two conditions, the two parties can communicate in the information-theoretically secure way, using the one-time pad [5].

4.3.2 The General Procedures of the QKD

In what follows, "Alice" and "Bob" denote the two parties that carry out a quantum key distribution protocol, and "Eve" denotes the eavesdropper, following the convention of the information theory. The common procedures of QKD protocols are given as follows.

Setups: Alice and Bob share the quantum channel in which they transmit quantum states as well as the authenticated public classical channel for the announcement. Eve can perform arbitrary attacks allowed by the law of quantum mechanics in the quantum channel and listen to all the announcements in the public channel.

Protocol:

1. Quantum communication:
   Alice and Bob encode classical information onto quantum states, send them through the quantum channel, and perform measurement on the quantum states they receive to read out the encoded information. Eve may perform arbitrary attacks in the quantum channel.
2. Sifting:
   Alice and Bob generate the binary strings z_A and z_B, which are selected or "sifted" from measurement outcomes according to the conditions defined in each protocol. z_A and z_B are called the sifted keys of Alice and Bob, respectively.
3. Information reconciliation:
   The obtained sifted keys z_A and z_B do not coincide in general due to noises or Eve's attacks in the quantum channel. Therefore, they have to correct the errors in their sifted keys. One way is that Alice sends the syndrome bits of her sifted key encrypted by a pre-shared secret key, and Bob corrects the errors on his sifted key z_B according to it. As a result, Bob obtains the reconciled key z_B^{rec}. This is called the "direct reconciliation". Alternatively, we can exchange the role of Alice and Bob in the above procedure. This alternative is called the "reverse reconciliation".
4. Privacy amplification:
   Eve may have part of the information of the (reconciled) keys z_A and z_B^{rec}. Alice and Bob estimate the amount of information leakage. Depending on the estimated leakage, they generate shorter keys z_A^{fin} and z_B^{fin} by acting a suitable linear hash function on z_A and z_B^{rec} (in the case of the direct reconciliation). These are the final keys.

Note that z_A, z_B, z_B^{rec}, z_A^{fin}, and z_B^{fin} are actually random variables and thus denoted with the symbol $\check{}$ in the following. Various types of QKD protocols mainly differ in the first line, i.e., the method of quantum communication. The sifting, the information reconciliation, and the privacy amplification after the quantum communication are referred to as (classical) post-processing. In most cases, the information reconciliation procedure does not differ so much among different QKD protocols. On the other hand, the privacy amplification procedure, or more precisely, the necessary amount of shortening the sifted key depends on which protocol to use. Thus, the development of the new QKD protocols requires both the procedure of quantum communication and the evaluation of the amount of privacy amplification.

4.3.3 The Security Condition of the Key in the QKD

In this section, the definition of security in the QKD is given. Let $\mathcal{H}_{ABE} := \mathcal{H}_A \otimes \mathcal{H}_B \otimes \mathcal{H}_E$ be the Hilbert space of the joint system of Alice's and Bob's final keys and Eve's quantum system. Let $\Pr(\cdot)$ be the probability distribution for all the random variables defined in the protocol and $\check{N}_{\text{fin}}$ be the length of the final key. Let $\rho^{\text{fin}}_{ABE|N} \in \mathfrak{D}(\mathcal{H}_{ABE})$ be the density operator of the state of Alice's and Bob's N-bit final keys as well as Eve's quantum state after all the attacks, given $\check{N}_{\text{fin}} = N$. When Alice and Bob abort the protocol, the final key length $\check{N}_{\text{fin}}$ is set to be zero. Then the definition of the ε-security is given by

$$\sum_{N\geq 1} \Pr(\check{N}_{\text{fin}} = N)\, d(\rho^{\text{fin}}_{ABE|N}, \rho^{\text{ideal}}_{ABE|N}) \leq \varepsilon, \tag{4.60}$$

where $d(\cdot,\cdot)$ denotes the trace distance in Definition 2.2.3, and

$$\rho^{\text{ideal}}_{ABE|N} := \sum_{z\in\{0,1\}^N} \frac{1}{2^N} |zz\rangle\langle zz|_{AB} \otimes \mathrm{Tr}_{AB}\left[\rho^{\text{fin}}_{ABE|N}\right]. \tag{4.61}$$

The parameter ε is an arbitrary positive constant and is referred to as the security parameter. This condition means that the final keys obtained in the protocol are, measured by the trace distance, ε-close to the ideal secret keys in which Alice's and Bob's keys perfectly coincide and are completely random for Eve. The trace distance is used in the security condition to ensure the monotonicity and the composability [60]. The monotonicity is necessary because the distinguishability between the final state of the protocol and the ideal state should not increase under quantum channels, i.e., CPTP maps. Since the trace distance is a CPTP monotone as shown in Proposition 2.2.5, the condition is satisfied. The composability is the requirement that, if Alice and Bob use an ε-secure key to perform an ε'-secure cryptographic protocol, the composite protocol should be at least $(\varepsilon + \varepsilon')$-secure. The composability is necessary because the security of a complex composite cryptographic protocol should be ensured by the security of its component protocols. In QKD, the composability follows from the triangle inequality (Corollary 2.2.4) of the trace distance.

4.3.4 An Approach to Prove the Security Condition

There are several approaches to prove the security of the QKD [49, 61–66]. One of the approaches for the security proof is to use the leftover hash lemma [49, 65], which utilizes the information-theoretic property of the universal$_2$ hash function [67, 68]. In this approach, we aim to evaluate the ε-smooth conditional min-entropy in Definition 4.2.8 of Alice's or Bob's sifted key conditioned on the adversary Eve. An advantage of this approach is the applicability to protocols in which the sifted keys are not binary numbers; the Gaussian-modulation continuous-variable QKD is an example.

Another approach, which is the main topic in this thesis, proves the security using complementarity between the bit (Z) and phase (X) bases [64]. By explicitly constructing a virtual procedure for correcting the phase errors of Alice's or Bob's sifted key that should be compatible with the actual protocol, we can ensure that the obtained final key is secret to Eve. Furthermore, if we find an optimal phase-error-correction procedure, the resulting key gain is optimal for the protocol [65, 66]. In this section, we review the approach in Refs. [64, 69] with a slight generalization. Without loss of generality, we here treat the direct reconciliation scenario. The case of reverse reconciliation can be treated similarly.

As shown in [64], the condition (4.60) can be divided into the following two conditions by applying triangle inequality to the trace distance. The first condition, the ε_{cor}-correctness condition, is the following:

$$\Pr(\check{z}_A^{\text{fin}} \neq \check{z}_B^{\text{fin}}) = \sum_{N \geq 1} \Pr(\check{N}_{\text{fin}} = N) \sum_{z,z' \in \{0,1\}^N, z \neq z'} \langle zz'|_{AB}\, \rho_{AB|N}^{\text{fin}}\, |zz'\rangle_{AB} \leq \varepsilon_{\text{cor}}, \tag{4.62}$$

where $\rho_{AB|N}^{\text{fin}} := \text{Tr}_E[\rho_{ABE|N}^{\text{fin}}]$. Equation (4.62) means that the probability of disagreement of Alice's and Bob's final keys is required to be no larger than ε_{cor}. The second condition, the ε_{sec}-secrecy condition, is defined as follows:

$$\sum_{N \geq 1} \Pr(\check{N}_{\text{fin}} = N)\, d(\rho_{AE|N}^{\text{fin}}, \rho_{AE|N}^{\text{ideal}}) \leq \varepsilon_{\text{sec}}, \tag{4.63}$$

where $\rho_{AE|N}^{\text{fin}} := \text{Tr}_B[\rho_{ABE|N}^{\text{fin}}]$ and $\rho_{AE|N}^{\text{ideal}} := \text{Tr}_B[\rho_{ABE|N}^{\text{ideal}}]$. With these two conditions, the protocol is ensured to be $(\varepsilon_{\text{cor}} + \varepsilon_{\text{sec}})$-secure.

The ε_{cor}-correctness condition is ensured if the failure probability of the information reconciliation is no larger than ε_{cor}; i.e., the reconciled keys $\check{z}_A$ and $\check{z}_B^{\text{rec.}}$ satisfy the following:

$$\begin{aligned}
&\Pr(\check{N}_{\mathrm{fin}} \geq 1, \check{z}_A \neq \check{z}_B^{\mathrm{rec}}) \\
&= \sum_{N' \geq 1} \sum_{z,z' \in \{0,1\}^{N'},\, z \neq z'} \Pr(\check{N}_{\mathrm{fin}} \geq 1, \check{N}_{\mathrm{sift}} = N', \check{z}_A = z, \check{z}_B^{\mathrm{rec}} = z') \qquad (4.64) \\
&\leq \varepsilon_{\mathrm{cor}}, \qquad (4.65)
\end{aligned}$$

where $\check{N}_{\mathrm{sift}}$ is the length of the sifted key or equivalently the length of the reconciled key. In order for this condition to be satisfied, the protocol has to be aborted with a high probability when the reconciliation fails.

In order to prove the $\varepsilon_{\mathrm{sec}}$-secrecy condition along with [64], a compatible virtual protocol is constructed as follows. First, $\check{N}_{\mathrm{sift}}$ (virtual) register qubits are introduced for Alice, and Alice's procedure of determining each bit of her sifted key in the actual protocol is replaced with the Z-basis measurement. The measurement is then postponed by replacing the privacy amplification procedure with the corresponding quantum operations as shown in Sect. 4.2.2. Thus Alice's $\check{N}_{\mathrm{fin}}$-bit final key is generated by the Z-basis measurements on $\check{N}_{\mathrm{fin}}$ qubits among Alice's register qubits in the virtual protocol. In constructing such a virtual protocol, Alice (and Bob) can do arbitrary quantum operations in the virtual protocol before the Z-basis measurements as long as the following condition is satisfied; for any attack by Eve in the actual protocol, there exists a corresponding attack in the virtual protocol such that the state on Alice's final key obtained by the Z-basis measurements and Eve's system is the same as that of the actual protocol; i.e., the probability $\Pr(\check{N}_{\mathrm{fin}} = N)$ and $\rho^{\mathrm{fin}}_{AE|N}$ are the same as the actual protocol. The condition for $\rho^{\mathrm{fin}}_{AE|N}$ can be rephrased as follows. Given $\check{N}_{\mathrm{fin}} = N$, let $\rho^{\mathrm{virt}}_{AE|N}$ be a quantum state on Alice's N register qubits and Eve's system just before Alice's Z-basis measurement in the virtual protocol. Let $\mathcal{E}^{\mathrm{meas}}_{A|N}$ be a CPTP map of Alice's Z-basis measurement in the virtual protocol given by

$$\mathcal{E}^{\mathrm{meas}}_{A|N}(\rho) = \sum_{z \in \{0,1\}^N} \langle z|\rho|z\rangle \, |z\rangle \langle z| \,. \qquad (4.66)$$

Then, they must satisfy $\mathcal{E}^{\mathrm{meas}}_{A|N} \otimes \mathrm{Id}_E(\rho^{\mathrm{virt}}_{AE|N}) = \rho^{\mathrm{fin}}_{AE|N}$ for Eve's corresponding attack in the virtual protocol. With $\mathcal{E}^{\mathrm{meas}}_{A|N}$ and $\rho^{\mathrm{virt}}_{AE|N}$, the secrecy condition (4.63) can be rewritten as follows:

$$\sum_{N \geq 1} \Pr(\check{N}_{\mathrm{fin}} = N)\, d\big(\mathcal{E}^{\mathrm{meas}}_{A|N} \otimes \mathrm{Id}_E(\rho^{\mathrm{virt}}_{AE|N}), \mathcal{E}^{\mathrm{meas}}_{A|N}\big(|0_X\rangle\langle 0_X|_A^{\otimes N}\big) \otimes \rho^{\mathrm{fin}}_{E|N}\big) \leq \varepsilon_{\mathrm{sec}}, \qquad (4.67)$$

where $\rho^{\mathrm{fin}}_{E|N} := \mathrm{Tr}_A\Big[\rho^{\mathrm{fin}}_{AE|N}\Big]$.

Now we find a sufficient condition in order for Eq. (4.67) to hold. The left-hand side of Eq. (4.67) is bounded from above as follows:

$$\begin{aligned}
&(\text{L.H.S. of Eq. (4.67)}) && (4.68)\\
&\le \sum_{N\ge 1} \Pr(\check{N}_{\mathrm{fin}} = N)\, d\big(\rho^{\mathrm{virt}}_{AE|N}, |0_X\rangle\langle 0_X|_A^{\otimes N} \otimes \rho^{\mathrm{fin}}_{E|N}\big) && (4.69)\\
&\le \sum_{N\ge 1} \Pr(\check{N}_{\mathrm{fin}} = N)\sqrt{1 - F\Big(\rho^{\mathrm{virt}}_{AE|N}, |0_X\rangle\langle 0_X|_A^{\otimes N} \otimes \rho^{\mathrm{fin}}_{E|N}\Big)} && (4.70)\\
&\le \Pr(\check{N}_{\mathrm{fin}} = 0)\sqrt{1-1^2} + \sum_{N\ge 1} \Pr(\check{N}_{\mathrm{fin}} = N)\sqrt{1 - \Big[F\Big(\rho^{\mathrm{virt}}_{A|N}, |0_X\rangle\langle 0_X|_A^{\otimes N}\Big)\Big]^2} && (4.71)\\
&\le \sqrt{1 - \Big[\Pr(\check{N}_{\mathrm{fin}} = 0) + \sum_{N\ge 1} \Pr(\check{N}_{\mathrm{fin}} = N)\, F\Big(\rho^{\mathrm{virt}}_{A|N}, |0_X\rangle\langle 0_X|_A^{\otimes N}\Big)\Big]^2} && (4.72)\\
&= \sqrt{1 - \Big[1 - \sum_{N\ge 1} \Pr(\check{N}_{\mathrm{fin}} = N)\, \Big(1 - \langle 0_X|_A^{\otimes N}\, \rho^{\mathrm{virt}}_{A|N}\, |0_X\rangle_A^{\otimes N}\Big)\Big]^2} && (4.73)
\end{aligned}$$

where $\rho^{\mathrm{virt}}_{A|N} := \mathrm{Tr}_E\Big[\rho^{\mathrm{virt}}_{AE|N}\Big]$. Here, the first inequality follows from Proposition 2.2.5, the second inequality follows from Proposition 2.2.12, the third inequality follows from Lemma 2.2.11, and the last inequality follows from the concavity of the function $f(x) = \sqrt{1-x^2}$. If

$$\sum_{N\ge 1} \Pr(\check{N}_{\mathrm{fin}} = N)\, \big(1 - \langle 0_X|_A^{\otimes N}\, \rho^{\mathrm{virt}}_{A|N}\, |0_X\rangle_A^{\otimes N}\big) \le \eta' \tag{4.74}$$

is satisfied for a (small) parameter η', then the left-hand side of Eq. (4.63) (or equivalently Eq. (4.67)) is bounded from above by

$$\sum_{N\ge 1} \Pr(\check{N}_{\mathrm{fin}} = N)\, d\big(\rho^{\mathrm{fin}}_{AE|N}, \rho^{\mathrm{ideal}}_{AE|N}\big) \le \sqrt{1-(1-\eta')^2} \le \sqrt{2\eta'}, \tag{4.75}$$

and thus the protocol is $\varepsilon_{\mathrm{sec}} = \sqrt{2\eta'}$-secret. Furthermore, the inequality (4.74) means that, if Alice measured $\rho^{\mathrm{virt}}_{A|N}$ on the X bases and obtained a (random) sequence $\check{\boldsymbol{x}}^{\mathrm{fin}}_A$, it satisfies

$$\Pr(\check{N}_{\mathrm{fin}} \ge 1, \check{\boldsymbol{x}}^{\mathrm{fin}}_A \ne \mathbf{0}) \le \eta'. \tag{4.76}$$

In other words, it suffices to show that Alice succeeded in correcting the "phase errors" on her register qubits with high probability. With these arguments, the privacy amplification can be regarded as the virtual phase-error-correction procedure on Alice's register qubits.

To sum up, in order to ensure the $(\varepsilon_{\mathrm{sec}} + \varepsilon_{\mathrm{cor}})$-security of the final key defined in Eq. (4.60), the following two conditions have to be satisfied. The first is the $\varepsilon_{\mathrm{cor}}$-correctness condition (4.62); Alice's sifted key $\check{z}_A$ and Bob's reconciled key $\check{z}^{\mathrm{rec}}_B$ after the information reconciliation coincides with probability no smaller than $1 - \varepsilon_{\mathrm{cor}}$, i.e.,

$$\Pr(\check{N}_{\text{fin}} \geq 1, \check{z}_A \neq \check{z}_B^{\text{rec}}) \leq \varepsilon_{\text{cor}} \tag{4.77}$$

The second is the ε_{sec}-secrecy condition (4.63). This condition can be replaced with the condition (4.67) by constructing a compatible virtual protocol that satisfies the followings. For any attack by Eve in the actual protocol, there exists a corresponding attack by Eve in the virtual protocol and Alice's quantum operations that result in the state $\rho_{AE|N}^{\text{virt}}$ satisfying

$$\mathcal{E}_{A|N}^{\text{meas}} \otimes \text{Id}_E \left(\rho_{AE|N}^{\text{virt}}\right) = \rho_{AE|N}^{\text{fin}} \tag{4.78}$$

and

$$\sum_{N\geq 1} \Pr(\check{N}_{\text{fin}} = N) \left(1 - \langle 0_X|_A^{\otimes N} \rho_{A|N}^{\text{virt}} |0_X\rangle_A^{\otimes N}\right) \leq \frac{\varepsilon_{\text{sec}}^2}{2} (= \eta') \tag{4.79}$$

at the same time, where $\mathcal{E}_{A|N}^{\text{meas}}$ is defined in Eq. (4.66). Note that the second condition (4.79) is equivalent to Eq. (4.76).

4.3.5 *The Privacy Amplification Using Dual Universal$_2$ Hashing*

In this section, the privacy amplification using dual universal$_2$ hashing is reviewed [68, 69]. Let N and N' be the length of the final key and the sifted key, respectively. For the privacy amplification, multiplying the (randomly generated) $N' \times N$ matrix $\check{G}$ on the sifted key $\check{z}_A$ to obtain the final key $\check{z}_A^{\text{fin}} = \check{z}_A \check{G}$ in the actual protocol is replaced with acting the unitary $U(\check{C})$ of Eq. (4.5) on Alice's register qubits followed by the Z-basis measurements $\mathcal{E}_{A|N}^{\text{meas}}$ on the first N qubits of the N'-qubit register. Here, $N' \times N'$ full-rank matrix $\check{C}$ satisfies $\check{G} = \check{C}(\boldsymbol{I}_N\ \boldsymbol{O})^\top$, where $\boldsymbol{I}_n$ denotes the $n \times n$ identity matrix and $\boldsymbol{O}$ denotes a zero matrix with the appropriate size. Let $\rho_{AE|N'}$ be the state of Alice's N'-qubit register and Eve's system just before the privacy amplification in the virtual protocol. Then for $N \geq 1$, we set

$$\rho_{AE|N}^{\text{virt}} = \sum_{\boldsymbol{y}\in\{0,1\}^{N'-N}} U_X^{\oplus}(\boldsymbol{v}(\boldsymbol{y}))\text{Tr}_{N'\setminus N}\left[\left(I_N \otimes |\boldsymbol{y}_X\rangle\langle \boldsymbol{y}_X|\right) U(\check{C})\rho_{AE|N'}U(\check{C})^\dagger\right]U_X^{\oplus}(\boldsymbol{v}(\boldsymbol{y}))^\dagger, \tag{4.80}$$

where $U_X^{\oplus}(\boldsymbol{v})$ is defined in Sect. 4.2.2, and all the unitaries act on Alice's N'-qubit register while act as identity on Eve's system. The state $\rho_{AE|N}^{\text{virt}}$ in Eq. (4.80) satisfies the condition (4.78) since $U_X^{\oplus}(\boldsymbol{v})$ does not change the Z-basis values of Alice's N'-qubit register, i.e.,

$$\mathcal{E}_{A|N}^{\text{meas}} \otimes \text{Id}_E \left(\rho_{AE|N}^{\text{virt}}\right) = \mathcal{E}_{A|N}^{\text{meas}} \otimes \text{Id}_E \left(\text{Tr}_{N'\setminus N}\left[U(\check{C})\rho_{AE|N'}U(\check{C})^\dagger\right]\right). \tag{4.81}$$

The role of $U_X^{\oplus}(\boldsymbol{v})$ as well as the $\boldsymbol{y}$ dependency of $\boldsymbol{v}$ in Eq. (4.80) will be revealed in the following. At this point, we recall the condition (4.79) or equivalently (4.76) for the secrecy. Suppose we performed X-basis measurement on $\rho_{A|N'} := \mathrm{Tr}_E[\rho_{AE|N'}]$ and obtained a random sequence $\check{\boldsymbol{x}}_A$. Then from Eqs. (4.5) and (4.80), it is related to the random sequence $\check{\boldsymbol{x}}_A^{\mathrm{fin}}$ in Eq. (4.76) with $\rho_{A|N'}$ not measured on the X basis by

$$\check{\boldsymbol{x}}_A^{\mathrm{fin}} = \check{\boldsymbol{x}}_A \check{H}' + \boldsymbol{v}(\check{\boldsymbol{x}}_A \check{H}), \tag{4.82}$$

where the matrices $\check{H}'$ and $\check{H}$ are defined as

$$\check{H}' := (\check{C}^{\top})^{-1}(\boldsymbol{I}_N \ \boldsymbol{O})^{\top}, \tag{4.83}$$

$$\check{H} := (\check{C}^{\top})^{-1}(\boldsymbol{O} \ \boldsymbol{I}_{N'-N})^{\top}. \tag{4.84}$$

Then, the condition (4.76) can be reinterpreted as

$$\Pr(\check{N}_{\mathrm{fin}} \geq 1, \check{\boldsymbol{x}}_A \check{H}' \neq \boldsymbol{v}(\check{\boldsymbol{x}}_A \check{H})) \leq \eta'. \tag{4.85}$$

We set $\boldsymbol{v}$ as follows. Given the $(N' - N)$-bit sequence $\check{\boldsymbol{x}}_A \check{H}$ and other random variables $\check{\boldsymbol{\zeta}}$ that can be defined in the virtual protocol (but not necessarily be observed in the actual protocol), we determine an estimate $\boldsymbol{x}_A^*(N', \check{\boldsymbol{\zeta}}, \check{\boldsymbol{x}}_A \check{H})$ of the sequence $\check{\boldsymbol{x}}_A$ and set $\boldsymbol{v} = \boldsymbol{x}_A^*(N', \check{\boldsymbol{\zeta}}, \check{\boldsymbol{x}}_A \check{H})\check{H}'$. (Thus, $\boldsymbol{v}$ also depends on N' and $\check{\boldsymbol{\zeta}}$.) Then, the condition (4.85) can be ensured if the following condition holds:

$$\Pr(\check{N}_{\mathrm{fin}} \geq 1, \check{\boldsymbol{x}}_A \neq \boldsymbol{x}_A^*(\check{N}_{\mathrm{sift}}, \check{\boldsymbol{\zeta}}, \check{\boldsymbol{x}}_A \check{H})) \leq \eta'. \tag{4.86}$$

This condition is nothing but the identifiability of the random sequence $\check{\boldsymbol{x}}_A$, given the $(N' - N)$-bit syndrome $\check{\boldsymbol{x}}_A \check{H}$ with the hash function $\check{H}$ and other random variables $\check{\boldsymbol{\zeta}}$ that can be defined in the virtual protocol. As the next step, suppose that, given $\check{N}_{\mathrm{sift}}$ and $\check{\boldsymbol{\zeta}}$, we can find a set $\mathcal{T}(\check{N}_{\mathrm{sift}}, \check{\boldsymbol{\zeta}})$ of candidates of the sequence $\check{\boldsymbol{x}}_A$ so that $\check{\boldsymbol{x}}_A$ belongs to $\mathcal{T}(\check{N}_{\mathrm{sift}}, \check{\boldsymbol{\zeta}})$ with a high probability, i.e.,

$$\Pr(\check{N}_{\mathrm{sift}} \geq 1, \check{\boldsymbol{x}}_A \notin \mathcal{T}(\check{N}_{\mathrm{sift}}, \check{\boldsymbol{\zeta}})) \leq \eta. \tag{4.87}$$

In general, the random variables $\check{\boldsymbol{\zeta}}$ consist of ones that can be observed in the actual protocol denoted by $\check{\boldsymbol{\alpha}}$ and ones that cannot be in the actual protocol denoted by $\check{\boldsymbol{\beta}}$. Thus we write $\check{\boldsymbol{\zeta}} = (\check{\boldsymbol{\alpha}}, \check{\boldsymbol{\beta}})$. As will be shown later, the required amount of the privacy amplification is determined by the cardinality $|\mathcal{T}(\check{N}_{\mathrm{sift}}, \check{\boldsymbol{\zeta}})|$ of the set $\mathcal{T}(\check{N}_{\mathrm{sift}}, \check{\boldsymbol{\zeta}})$. Since we need to know the amount of the privacy amplification in the actual protocol, we define an integer-valued function $N_{\mathrm{PA}}(\check{N}_{\mathrm{sift}}, \check{\boldsymbol{\alpha}})$ that depends on $\check{\boldsymbol{\alpha}}$ of $\check{\boldsymbol{\zeta}}$ and satisfies the following:

$$\Pr(\log |\mathcal{T}(\check{N}_{\mathrm{sift}}, \check{\boldsymbol{\zeta}})| \leq N_{\mathrm{PA}}(\check{N}_{\mathrm{sift}}, \check{\boldsymbol{\alpha}}) \mid \check{N}_{\mathrm{sift}} \geq 1) = 1. \tag{4.88}$$

Suppose further that for each final key length N and sifted key length N', the $N' \times N$ matrix $\check{H}$ is randomly chosen from the universal$_2$ hash function family; i.e., the following holds [49, 67, 68]:

$$\forall \boldsymbol{x}, \boldsymbol{x}' \in \{0,1\}^{N'}, \boldsymbol{x} \neq \boldsymbol{x}', \qquad \Pr(\boldsymbol{x}\check{H} = \boldsymbol{x}'\check{H}) \leq 2^{-(N'-N)}. \tag{4.89}$$

We then construct $\boldsymbol{x}_A^*(\check{N}_{\text{sift}}, \check{\zeta}, \check{\boldsymbol{x}}_A\check{H})$ as follows. Given the syndrome $\check{\boldsymbol{x}}_A\check{H}$, if there is only one element in $\mathcal{T}(\check{N}_{\text{sift}}, \check{\zeta})$ that is consistent with the syndrome $\check{\boldsymbol{x}}_A\check{H}$, then that element is chosen as $\boldsymbol{x}_A^*(\check{N}_{\text{sift}}, \check{\zeta}, \check{\boldsymbol{x}}_A\check{H})$. Otherwise, an arbitrary sequence is chosen as $\boldsymbol{x}_A^*(\check{N}_{\text{sift}}, \check{\zeta}, \check{\boldsymbol{x}}_A\check{H})$. With this construction and in the case $\check{N}_{\text{sift}} = N'$, $\check{N}_{\text{fin}} = N$, $\check{\zeta} = \zeta' = (\alpha', \beta')$, and $\check{\boldsymbol{x}}_A = \boldsymbol{x} \in \mathcal{T}(N', \zeta')$, the condition $\boldsymbol{x}_A^*(N', \zeta', \boldsymbol{x}\check{H}) = \boldsymbol{x}$ holds if $\boldsymbol{x}\check{H} \neq \boldsymbol{y}\check{H}$ holds for any $\boldsymbol{y} \in \mathcal{T}(N', \zeta') \setminus \{\boldsymbol{x}\}$. Combining this with Eq. (4.88), we have, for $\Pr(\check{N}_{\text{sift}} = N', \check{\zeta} = \zeta') > 0$ and $\boldsymbol{x} \in \mathcal{T}(N', \zeta')$,

$$\Pr(\boldsymbol{x} \neq \boldsymbol{x}_A^*(N', \zeta', \boldsymbol{x}\check{H})) \leq 2^{N_{\text{PA}}(N', \alpha') - (N'-N)}. \tag{4.90}$$

Then, unless the protocol is aborted before the privacy amplification step, we set

$$\check{N}_{\text{fin}} = \max\{\check{N}_{\text{sift}} - N_{\text{PA}}(\check{N}_{\text{sift}}, \check{\alpha}) - s, 0\}, \tag{4.91}$$

which leads to

$$\Pr(\check{N}_{\text{fin}} = \check{N}_{\text{sift}} - N_{\text{PA}}(\check{N}_{\text{sift}}, \check{\alpha}) - s \mid \check{N}_{\text{fin}} \geq 1) = 1. \tag{4.92}$$

Combining this with Eqs. (4.87) and (4.90), we have

$$\Pr(\check{N}_{\text{fin}} \geq 1, \check{\boldsymbol{x}}_A \neq \boldsymbol{x}_A^*(\check{N}_{\text{sift}}, \check{\zeta}, \check{\boldsymbol{x}}_A\check{H})) \tag{4.93}$$

$$\leq \Pr\Big(\check{N}_{\text{fin}} \geq 1, \check{\boldsymbol{x}}_A \notin \mathcal{T}(\check{N}_{\text{sift}}, \check{\zeta}) \cup \Big[\check{\boldsymbol{x}}_A \in \mathcal{T}(\check{N}_{\text{sift}}, \check{\zeta}) \cap \check{\boldsymbol{x}}_A \neq \boldsymbol{x}_A^*(\check{N}_{\text{sift}}, \check{\zeta}, \check{\boldsymbol{x}}_A\check{H})\Big]\Big) \tag{4.94}$$

$$\leq \eta + 2^{-s}. \tag{4.95}$$

Setting $\eta' = \eta + 2^{-s}$ leads to Eq. (4.86) and thus Eq. (4.76). Finally, it is known that randomly choosing $\check{H}$ from the universal$_2$ hash function family amounts to randomly choosing $\check{G}$ from the dual universal$_2$ hash function family [68]. (These two relate through the matrix $\check{C}$.) Therefore, we perform the dual universal$_2$ hashing in the privacy amplification step of the actual protocol in order for the argument in this section to hold. Note that $\check{H}$ can be chosen from the almost universal$_2$ hash function families, which is more general than the universal$_2$ hash function families, at the cost of the additional amount of the syndrome extraction in the virtual protocol. In this case, the almost dual universal$_2$ hashing will be performed in the actual protocol at the cost of the additional amount of the privacy amplification. See Ref. [68] for more details.

To sum up, if all of the following conditions are satisfied, the privacy amplification using the dual universal$_2$ hashing ensures ε_{sec}-secrecy with $\varepsilon_{\text{sec}} = \sqrt{2(\eta + 2^{-s})}$.

- Given $\check{N}_{\text{sift}} \geq 1$, let $\check{\boldsymbol{x}}_A$ be the outcome of the X-basis measurement on Alice's $\check{N}_{\text{sift}}$-qubit register just before the privacy amplification step in the virtual protocol. There exists a set-valued function $\mathcal{T}(\check{N}_{\text{sift}}, \check{\boldsymbol{\zeta}})$ and an integer-valued function $N_{\text{PA}}(\check{N}_{\text{sift}}, \check{\boldsymbol{\alpha}})$ of the sifted-key length $\check{N}_{\text{sift}}$ and other random variables $\check{\boldsymbol{\zeta}} = (\check{\boldsymbol{\alpha}}, \check{\boldsymbol{\beta}})$ defined in the virtual protocol, where $\check{\boldsymbol{\alpha}}$ is observable in the actual protocol, such that
$$\Pr(\check{N}_{\text{sift}} \geq 1, \check{\boldsymbol{x}}_A \notin \mathcal{T}(\check{N}_{\text{sift}}, \check{\boldsymbol{\zeta}})) \leq \eta, \tag{4.96}$$
and
$$\Pr(\log |\mathcal{T}(\check{N}_{\text{sift}}, \check{\boldsymbol{\zeta}})| \leq N_{\text{PA}}(\check{N}_{\text{sift}}, \check{\boldsymbol{\alpha}}) \mid \check{N}_{\text{sift}} \geq 1) = 1. \tag{4.97}$$
- Unless the protocol is aborted before the privacy amplification step, the final key length is chosen to be
$$\check{N}_{\text{fin}} = \max\{\check{N}_{\text{sift}} - N_{\text{PA}}(\check{N}_{\text{sift}}, \check{\boldsymbol{\alpha}}) - s, 0\} \tag{4.98}$$
at the privacy amplification step.

4.3.6 Key Rate of the QKD Protocol

In general, secret keys may be consumed in the QKD protocol in order to, for example, authenticate the announcement or send encrypted messages such as the bit error syndromes. Therefore, if the number of the consumed secret keys in the whole protocol is denoted by $\check{N}_{\text{KC}}$, the net key gain is given by $\check{N}_{\text{fin}} - \check{N}_{\text{KC}}$. If the total number of communication rounds between Alice and Bob is N_{tot}, then the key rate per pulse is defined as $(\check{N}_{\text{fin}} - \check{N}_{\text{KC}})/N_{\text{tot}}$. This quantity shows an efficiency for the key gain of the protocol and thus is the performance index for the QKD protocol. It is implied in the previous section that $(N_{\text{PA}}(\check{N}_{\text{sift}}, \check{\boldsymbol{\alpha}}) + s)$-bit privacy amplification is sufficient to ensure the secrecy condition when we use the dual universal$_2$ hashing. Therefore, the key rate can alternatively be given by
$$(\check{N}_{\text{sift}} - \check{N}_{\text{KC}} - N_{\text{PA}}(\check{N}_{\text{sift}}, \check{\boldsymbol{\alpha}}) - s)/N_{\text{tot}}. \tag{4.99}$$

4.4 Finite-Size Security of Continuous-Variable QKD with Digital Signal Processing

In this section, the composable security of a binary-modulation continuous-variable QKD protocol is proved against arbitrary attacks in the finite-size regime. The proposed security proof can be applied to the case in which the binned homodyne and heterodyne measurements are used (see also [70]). The key to our security proof is an estimation method we develop in Sect. 4.4.1 using the heterodyne measurement and classical post-processing, which is suited for analysis of confidence region in the finite-size regime. The outcome of the heterodyne measurement, which is unbounded, is converted to a bounded value by a smooth function such that its expectation is proved to be no larger than the fidelity of the input pulse to a coherent state. This allows us to use a standard technique to derive a lower bound on the fidelity with a required confidence level in the finite-size regime. The fidelity as a measure of disturbance in the binary modulation protocol is essentially the same as what is monitored through bit errors in the B92 protocol [6, 71, 72]. Using this similarity, we derive an operator inequality through which we construct a security proof based on a reduction to the distillation of entangled qubit pairs [61, 73], a technique frequently used for discrete-variable QKD protocols.

4.4.1 Estimation of the Fidelity to a Coherent State

We first introduce a test scheme to estimate the fidelity between an optical state ρ and the vacuum state $|0\rangle\langle 0|$ through a heterodyne measurement. For a state ρ of a single optical mode, the heterodyne measurement (with its POVM defined in Eq. (3.90)) produces an outcome $\check{\alpha} \in \mathbb{C}$ with a probability density

$$q_\rho(\alpha)\, d^2\alpha := \langle\alpha|\, \rho\, |\alpha\rangle \,\frac{d^2\alpha}{\pi}, \tag{4.100}$$

where the coherent state $|\alpha\rangle$ is defined in Eq. (3.85). We refer to the expectation associated with the distribution $q_\rho(\alpha)$ simply as $\mathbb{E}_\rho$. To construct a lower bound on the fidelity $\langle 0|\, \rho\, |0\rangle$ from $\check{\alpha}$, we will use the associated Laguerre polynomials which are given by

$$L_n^{(k)}(\nu) := (-1)^k \frac{d^k L_{n+k}(\nu)}{d\nu^k}, \tag{4.101}$$

where

$$L_n(\nu) := \frac{e^\nu}{n!}\frac{d^n}{d\nu^n}(e^{-\nu}\nu^n) \tag{4.102}$$

are the Laguerre polynomials. Our test scheme is based on the following theorem.

Theorem 4.4.1 *Let $\Lambda_{m,r}(\nu)$ $(\nu \geq 0)$ be a bounded function given by*

$$\Lambda_{m,r}(\nu) := e^{-r\nu}(1+r)L_m^{(1)}((1+r)\nu), \tag{4.103}$$

for an integer $m \geq 0$ and a real number $r > 0$. Then, for any density operator ρ, we have

$$\mathbb{E}_\rho[\Lambda_{m,r}(|\check{\alpha}|^2)] = \langle 0|\,\rho\,|0\rangle + \sum_{n=m+1}^{\infty} \frac{\langle n|\,\rho\,|n\rangle}{(1+r)^n} I_{n,m}, \tag{4.104}$$

where $I_{n,m}$ are constants satisfying $(-1)^m I_{n,m} > 0$.

Proof From Eq. (4.100), the expectation value of $\Lambda_{m,r}(|\check{\alpha}|^2)$ when given a measured state ρ is given by

$$\begin{aligned}
\mathbb{E}_\rho[\Lambda_{m,r}(|\check{\alpha}|^2)] &= \int_{\alpha\in\mathbb{C}} \Lambda_{m,r}\left(|\alpha|^2\right) q_\rho(\alpha)\, d^2\alpha \\
&= \int_0^\infty d\nu\, \Lambda_{m,r}(\nu) \left(\int_0^{2\pi} \frac{d\theta}{2\pi} \langle \sqrt{\nu}e^{i\theta}|\,\rho\,|\sqrt{\nu}e^{i\theta}\rangle \right) \\
&= \int_0^\infty d\nu\, \Lambda_{m,r}(\nu) \left(\sum_{n=0}^{\infty} \frac{\nu^n e^{-\nu}}{n!} \langle n|\,\rho\,|n\rangle \right) \\
&= \sum_{n=0}^{\infty} \frac{\langle n|\,\rho\,|n\rangle\, I_{n,m}}{(1+r)^n},
\end{aligned} \tag{4.105}$$

where

$$I_{n,m} := \frac{1}{n!} \int_0^\infty d\nu\, e^{-\nu} \nu^n L_m^{(1)}(\nu) \tag{4.106}$$

for integers $n, m \geq 0$.

The following three properties hold for $I_{n,m}$:

(i) $I_{n,m} = 0$ for $m \geq n \geq 1$.

This results from orthogonality relations of the associated Laguerre polynomials, that is,

$$\int_0^\infty L_n^{(1)}(\nu) L_m^{(1)}(\nu) \nu e^{-\nu}\, d\nu = (n+1)\delta_{n,m}. \tag{4.107}$$

Since the polynomial ν^{n-1} can be written as a linear combination of lower order polynomials $\{L_l^{(1)}(\nu)\}_{0\leq l\leq n-1}$, $I_{n,m}$ vanishes whenever $m \geq n \geq 1$.

(ii) $(-1)^m I_{n,m} > 0$ for $n > m \geq 0$.

This property is shown as follows. First, the associated Laguerre polynomials satisfy the following recurrence relation for $m \geq 1$ [74]:

$$m L_m^{(1)}(\nu) = \nu \frac{dL_m^{(1)}}{d\nu}(\nu) + (m+1) L_{m-1}^{(1)}(\nu). \tag{4.108}$$

Substituting this to Eq. (4.106) and using integration by parts, we have

$$I_{n,m} = \frac{n+m}{n} I_{n-1,m} - \frac{m+1}{n} I_{n-1,m-1}. \tag{4.109}$$

for $n \geq 1$ and $m \geq 1$. The property (ii) is then proved by induction over m. For $m = 0$, it is true since $I_{n,0} = 1 > 0$. When $(-1)^{m-1} I_{n,m-1} > 0$ for $n > m-1$, we can prove $(-1)^m I_{n,m} > 0$ for $n > m$ by using Eq. (4.109) recursively with $I_{m,m} = 0$ from the property (i).

(iii) $I_{0,m} = 1$ for $m \geq 0$.

This also follows from property (i) and Eq. (4.109) for $n = 1$ and $m \geq 1$, which leads to $I_{0,m} = I_{0,0} = 1$. Combining properties (i), (ii), and (iii) shows Eq. (4.104). □

Note that Eq. (4.104) can also be interpreted as an operator equality

$$\int_{\alpha \in \mathbb{C}} \frac{d^2\alpha}{\pi} \Lambda_{m,r}(|\alpha|^2) \, |\alpha\rangle\langle\alpha| = |0\rangle\langle 0| + \sum_{n=m+1}^{\infty} \frac{I_{n,m}}{(1+r)^n} |n\rangle\langle n| \,, \tag{4.110}$$

which converges σ-weakly.

As a corollary, we obtain the following.

Corollary 4.4.2 *Let $|\beta\rangle$ $(\beta \in \mathbb{C})$ be the coherent state with the amplitude β. Then, for any $\beta \in \mathbb{C}$ and for any odd positive integer m, we have*

$$\mathbb{E}_\rho[\Lambda_{m,r}(|\breve{\alpha} - \beta|^2)] \leq \langle\beta|\, \rho \,|\beta\rangle \,. \tag{4.111}$$

Proof From Eq. (4.104) of Theorem 4.4.1, for any odd positive integer m, we have

$$\mathbb{E}_\rho[\Lambda_{m,r}(|\breve{\alpha}|^2)] \leq \langle 0|\, \rho \,|0\rangle \,. \tag{4.112}$$

Let D_β be a displacement operator satisfying

$$D_\beta \,|0\rangle\langle 0|\, D_\beta^\dagger = |\beta\rangle\langle\beta| \,, \tag{4.113}$$

and $D_\beta^\dagger = D_{-\beta}$. With $\tilde{\rho} := D_\beta \rho D_\beta^\dagger$, we have $q_{\tilde{\rho}}(\alpha) = q_\rho(\alpha - \beta)$ for probability density function of heterodyne measurement outcome, which implies that

$$\begin{aligned} \mathbb{E}_{\tilde{\rho}}[\Lambda_{m,r}(|\breve{\alpha} - \beta|^2)] &= \mathbb{E}_\rho[\Lambda_{m,r}(|\breve{\alpha}|^2)] \\ &\leq \langle 0|\, \rho \,|0\rangle \\ &= \langle\beta|\, \tilde{\rho} \,|\beta\rangle \,. \end{aligned} \tag{4.114}$$

Since this holds for any $\tilde{\rho}$, we proved Eq. (4.111). □

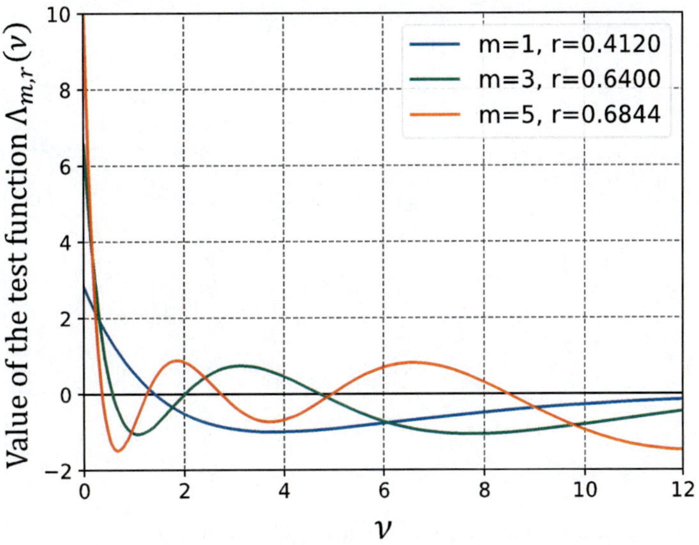

Fig. 4.2 Example of the test functions $\Lambda_{m,r}$ used in the estimation. The values of r in the figure are chosen so that the range of $\Lambda_{m,r}$ is minimized for given m. In general, the minimum range of the function $\Lambda_{m,r}$ becomes larger as m increases. The pair $(m, r) = (1, 0.4120)$ was used in the numerical simulation of key rates below

As seen in Fig. 4.2,[1] the absolute value and the slope of the function $\Lambda_{m,r}$ are moderate for small values of m and r, which is advantageous in executing the test in a finite duration with a finite resolution. Later we will see that the range of the function $\Lambda_{m,r}$ affects the speed of convergence to the expectation value in the case of a finite number of repetitions (i.e., a finite duration). Furthermore, when the outcome of the heterodyne measurement is digitized (i.e., has a finite resolution), assume that a digitized outcome $\check{\alpha}_{\mathrm{dig}}$ ensures that the true value $\check{\alpha}$ lies in a range $\Omega(\check{\alpha}_{\mathrm{dig}})$. Then, we need only to replace $\Lambda_{m,r}(|\check{\alpha} \pm \beta|^2)$ with its worst-case value, $\min\{\Lambda_{m,r}(|\check{\alpha} \pm \beta|^2) \mid \check{\alpha} \in \Omega(\check{\alpha}_{\mathrm{dig}})\}$, in order for the inequality (4.111) to be satisfied. As seen in Fig. 4.2, the slope of the function $\Lambda_{m,r}(\nu)$ is moderate and goes to zero for $\nu \to \infty$. This means that the worst-case value can be made close to the true value, leading to a small influence on the tightness of the estimation.

Compared to a similar method proposed in [75], our method excels in its tightness for weak input signals; we see from Eq. (4.104) that, regardless of the value of r, the inequality (4.112) saturates when ρ has at most m photons. This is crucial for the use in the QKD protocols in which tightness directly affects the efficiency of the key generation. Furthermore, from Eq. (4.104), the inequality (4.111) can be made arbitrarily tight by taking arbitrarily large m and r. Therefore, our estimation method is asymptotically sharp. See Ref. [76] for further generalizations.

[1] Modified from [46], Copyright 2021, The Author(s) licensed under CC BY 4.0.

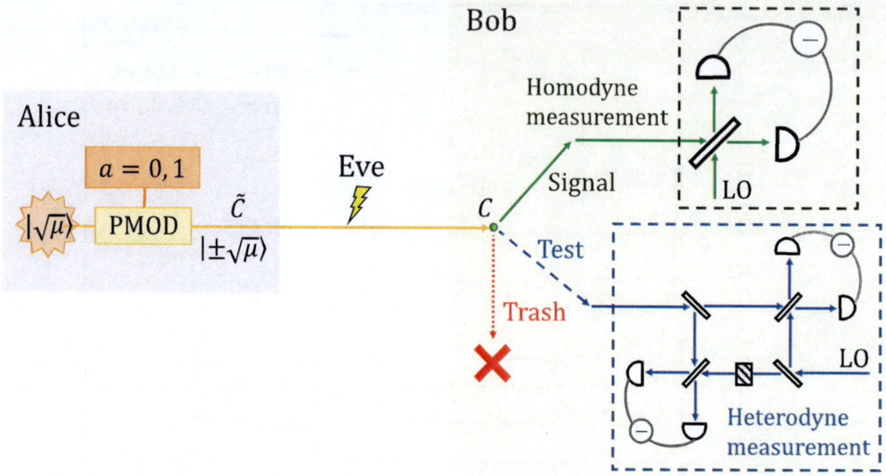

Fig. 4.3 The proposed continuous-variable quantum key distribution protocol. Alice generates a random bit $a \in \{0, 1\}$ and sends a coherent state with an amplitude $(-1)^a\sqrt{\mu}$. Bob chooses one of the three measurements based on the predetermined probability. In the signal round, Bob performs the homodyne measurement on the received optical pulse and obtains an outcome $\check{q}$. In the test round, Bob performs the heterodyne measurement on the received optical pulse and obtains an outcome $\check{\alpha}$. In the trash round, he produces no outcome

4.4.2 Proposed Protocol

Based on this fidelity test, we propose the following discrete-modulation QKD protocol (see Fig. 4.3).[2] Prior to the protocol, Alice and Bob determine the number of rounds N, the acceptance probability of the homodyne measurement $f_{\mathrm{suc}}(q)$ for $q \in \mathbb{R}$ satisfying $f_{\mathrm{suc}}(q) + f_{\mathrm{suc}}(-q) \leq 1$, the parameters for the test function (m, r) with m being positive odd integer and r being positive real, and the protocol parameters $(\mu, p_{\mathrm{sig}}, p_{\mathrm{test}}, p_{\mathrm{trash}}, \beta, s)$ with $p_{\mathrm{sig}} + p_{\mathrm{test}} + p_{\mathrm{trash}} = 1$, where all the parameters are positive. Alice and Bob then run the following protocol.

Actual protocol

1. Alice generates a random bit $a \in \{0, 1\}$ and sends an optical pulse $\tilde{C}$ in a coherent state with an amplitude $(-1)^a\sqrt{\mu}$ to Bob. She repeats it N times.
2. For each of the received N pulses, Bob chooses a label from {signal, test, trash} with probabilities p_{sig}, p_{test}, and p_{trash}, respectively. According to the label, Alice and Bob do one of the following procedures.

[signal] Bob performs the homodyne measurement (defined in Eq. (3.39)) on the received optical pulse C, and obtains an outcome $\check{q} \in \mathbb{R}$. Bob defines a bit b as $b = 0$ with a probability $f_{\mathrm{suc}}(\check{q})$ and $b = 1$ with a probability $f_{\mathrm{suc}}(-\check{q})$, and

[2] Modified from [46], Copyright 2021, The Author(s) licensed under CC BY 4.0.

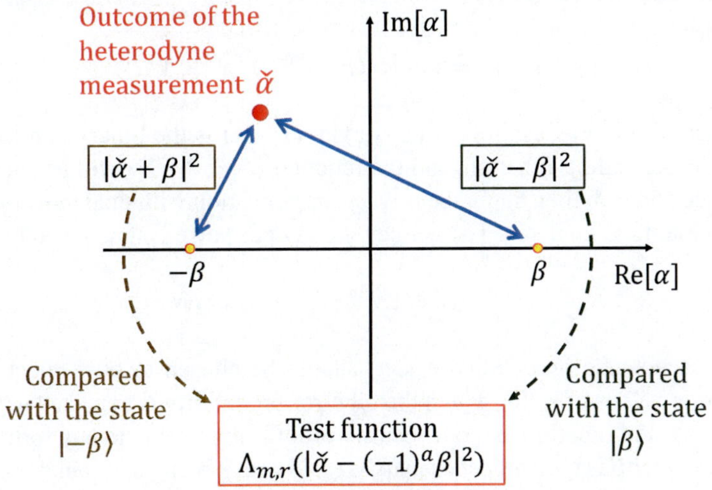

Fig. 4.4 A schematic description of the usage of obtained outcomes in the heterodyne measurement. In order to estimate the lower bound on the fidelity to the coherent states $|\pm\beta\rangle$, the squared distance between the outcome $\check{\alpha}$ and the objective point $(-1)^a\beta$ (i.e., $|\check{\alpha}-(-1)^a\beta|^2$) is put into the test function $\Lambda_{m,r}$

otherwise regards the round as "failure". The round in which Bob defines the bit b is regarded as "success". He announces success/failure of the detection. In the case of a success, Alice (resp. Bob) keeps a (b) as a sifted key bit.

[test] Bob performs the heterodyne measurement on the received optical pulse C, and obtains an outcome $\check{\alpha}$. Alice announces her bit a. Bob calculates the value of $\Lambda_{m,r}(|\check{\alpha}-(-1)^a\beta|^2)$. (See Fig. 4.4.)[3]

[trash] Alice and Bob produce no outcomes.

3. We refer to the numbers of "success" and "failure" signal rounds, test rounds, and trash rounds as $\check{N}^{\text{suc}}$, $\check{N}^{\text{fail}}$, $\check{N}^{\text{test}}$, and $\check{N}^{\text{trash}}$, respectively. ($N=\check{N}^{\text{suc}}+\check{N}^{\text{fail}}+\check{N}^{\text{test}}+\check{N}^{\text{trash}}$ holds by definition. Note also that the sifted key length of this protocol is equal to $\check{N}^{\text{suc}}$.) Bob calculates the sum of $\Lambda_{m,r}(|\check{\alpha}-(-1)^a\beta|^2)$ obtained in the $\check{N}^{\text{test}}$ test rounds, which is denoted by $\check{F}$.
4. For the information reconciliation, they use $(N_{\text{EC}}+s')$-bits of encrypted communication consuming a pre-shared secret key to do the following. Alice sends Bob an N_{EC}-bit syndrome of a linear code for her sifted key. Bob reconciles his sifted key accordingly. Alice and Bob verify the correction by comparing s' bits via universal$_2$ hashing [67].
5. Bob computes and announces the final key length $\check{N}^{\text{fin}}$ according to

$$\check{N}^{\text{fin}}=\max\left\{\check{N}^{\text{suc}}-\lceil\check{N}^{\text{suc}}h(\check{p})\rceil-s,0\right\} \tag{4.115}$$

[3] Modified from [46], Copyright 2021, The Author(s) licensed under CC BY 4.0.

with

$$\check{p} := \min\{U(\check{F}, \check{N}^{\text{trash}})/\check{N}^{\text{suc}}, 1/2\}, \tag{4.116}$$

where $h(x) := -x\log(x) - (1-x)\log(1-x)$ is the binary entropy function, $\lceil\cdot\rceil$ is the ceiling function, and the function $U(\check{F}, \check{N}^{\text{trash}})$ will be specified later. Alice and Bob then apply the privacy amplification with dual universal$_2$ hashing to obtain the final key. The net key gain $\check{G}$ per pulse is thus given by

$$\check{G} = (\check{N}^{\text{fin}} - N_{\text{EC}} - s')/N.$$

The acceptance probability $f_{\text{suc}}(q)$ should be chosen to post-select the rounds with larger values of q, for which the bit error probability is expected to be lower. It is ideally a step function, but our security proof is applicable to any form of $f_{\text{suc}}(q)$. The parameter β is typically chosen to be $\sqrt{\eta\mu}$ with η being a nominal transmissivity of the quantum channel, but the security proof itself holds for any choice of β. The parameters s and s' are related to the overall security parameter in the security proof below.

4.4.3 Security Proof

We prove the security of this protocol based on the approach summarized in Sect. 4.3; i.e., prove the ε_{cor}-correctness Eq. (4.77) and ε_{sec}-secrecy Eq. (4.63) of the protocol, which results in $(\varepsilon_{\text{cor}} + \varepsilon_{\text{sec}})$-security in Eq. (4.60). Due to the property of the universal$_2$ hashing [49, 67, 68] (see also Eq. (4.90)), our protocol achieves ε_{cor}-correctness with $\varepsilon_{\text{cor}} = 2^{-s'}$ via the verification in Step 4. In order to prove the ε_{sec}-secrecy, we determine a sufficient amount of the privacy amplification according to Shor and Preskill [73, 77], which can be regarded as a special case of the security proof summarized in Sect. 4.3.4. We consider a coherent version of Steps 1 and 2, in which Alice and Bob share an entangled pair of qubits for each success signal round, such that their Z-basis-measurement outcomes correspond to the sifted key bits a and b. For Alice, we introduce a qubit A and assume that she entangles it with an optical pulse $\tilde{C}$ in a state

$$|\Psi\rangle_{A\tilde{C}} := \frac{|0\rangle_A |\sqrt{\mu}\rangle_{\tilde{C}} + |1\rangle_A |-\sqrt{\mu}\rangle_{\tilde{C}}}{\sqrt{2}}. \tag{4.117}$$

Then, Step 1 is equivalent to the preparation of $|\Psi\rangle_{A\tilde{C}}$ followed by a measurement of the qubit A on Z basis $\{|0\rangle, |1\rangle\}$ to determine the bit value a. For Bob, we construct a process of probabilistically converting the received optical pulse C to a qubit B (See Fig. 4.5). Consider a completely positive (CP) map defined by

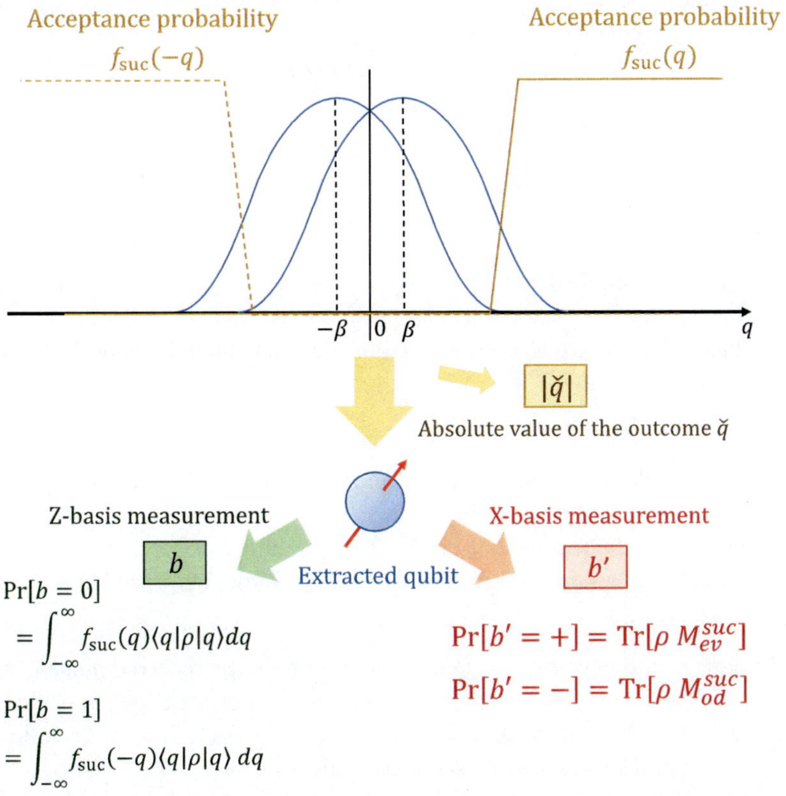

Fig. 4.5 Bob's qubit extraction in the entanglement-sharing protocol. Bob performs on the optical pulse a non-demolition projective measurement with which the absolute value $|q|$ of the outcome of the homodyne measurement $\check{q}$ is determined. Then, Bob prepares a qubit B in a state determined by the operation $(K^{(|q|)}(\cdot)K^{(|q|)\dagger} + K^{(-|q|)}(\cdot)K^{(-|q|)\dagger})dq$ with $K^{(|q|)}$ defined in Eq. (4.119). The Z-basis measurement on this qubit gives the same sifted key bit b as that in the actual protocol. On the other hand, the X-basis measurement on this qubit reveals the parity of photon number of the received optical pulse

$$\mathcal{F}_{C\to B}(\rho_C) := \int_{-\infty}^{\infty} dq\, K^{(q)}\rho_C K^{(q)\dagger} \tag{4.118}$$

with

$$K^{(q)} := \sqrt{f_{\rm suc}(q)}\left(|0\rangle_B\langle q|_C + |1\rangle_B\langle -q|_C\right), \tag{4.119}$$

where $\langle q|$ maps a state vector to the value of its wave function at q (see Eq. (3.8)). When the pulse C is in a state ρ_C, the corresponding process succeeds with a probability $p_{\rm suc}$ and then prepares the qubit B in a state ρ_B, where $p_{\rm suc}\rho_B = \mathcal{F}_{C\to B}(\rho_C)$. If the qubit B is further measured on Z basis, probabilities of the outcomes $b = 0, 1$ are given respectively by

$$p_{\text{suc}} \langle 0| \rho_B |0\rangle = \int_{-\infty}^{\infty} f_{\text{suc}}(q)dq \; \langle q| \rho_C |q\rangle , \tag{4.120}$$

$$p_{\text{suc}} \langle 1| \rho_B |1\rangle = \int_{-\infty}^{\infty} f_{\text{suc}}(q)dq \; \langle -q| \rho_C |-q\rangle , \tag{4.121}$$

which shows the equivalence to the signal round in Step 2. This is illustrated in[4] Fig. 4.5.

To clarify the above observation, we introduce an entanglement-sharing protocol defined in the following.

Entanglement-sharing protocol

1′. Alice prepares a qubit A and an optical pulse $\tilde{C}$ in a state $|\Psi\rangle_{A\tilde{C}}$ defined in (4.117). She repeats it N times.

2′. For each of the received N pulses, Bob announces a label in the same way as that in Step 2. Alice and Bob do one of the following procedures according to the label.

[signal] Bob performs a quantum operation on the received pulse C specified by the CP map $\mathcal{F}_{C\to B}$ to determine success/failure of detection and to obtain a qubit B upon success. He announces success/failure of detection. In the case of a success, Alice keeps her qubit A.

[test] Bob performs a heterodyne measurement on the received optical pulse C, and obtains an outcome $\check{\alpha}$. Alice measures her qubit A on the Z basis and announces the outcome $a \in \{0, 1\}$. Bob calculates the value of $\Lambda_{m,r}(|\check{\alpha} - (-1)^a\beta|^2)$.

[trash] Alice measures her qubit A on the X basis to obtain $a' \in \{+, -\}$, where $+ := 0_X$ and $- := 1_X$.

3′. $\check{N}^{\text{suc}}$, $\check{N}^{\text{fail}}$, $\check{N}^{\text{test}}$, $\check{N}^{\text{trash}}$, and $\check{F}$ are defined in the same way as those in Step 3. Let $\check{Q}_-$ be the number of rounds in the $\check{N}^{\text{trash}}$ trash rounds with $a' = -$.

This protocol leaves $\check{N}^{\text{suc}}$ pairs of qubits shared by Alice and Bob. If they measure these qubits on the Z bases to define the sifted key bits, the whole procedure is equivalent to Steps 1 through 3 of the actual protocol (see Fig. 4.6). Alice's measurements on the X basis $\{|\pm\rangle := (|0\rangle + |1\rangle)/\sqrt{2}\}$ in the trash rounds are added for later security argument, and they do not affect the equivalence.

The Shor-Preskill argument proceeds the security proof as follows. Suppose that, after the entanglement-sharing protocol, the Controlled-NOT operation CNOT is applied on each pair of qubits, where CNOT $:= |0\rangle\langle 0|_A \otimes I_B + |1\rangle\langle 1|_A \otimes \sigma_B^X$. Alice then performs the quantum-mechanical version of the privacy amplification in Step 5 of the actual protocol as in Sect. 4.2.2 and successively measures the qubits on the Z bases to obtain the final key. Since CNOT does not affect the Z-basis value of

[4] Modified from [46], Copyright 2021, The Author(s) licensed under CC BY 4.0.

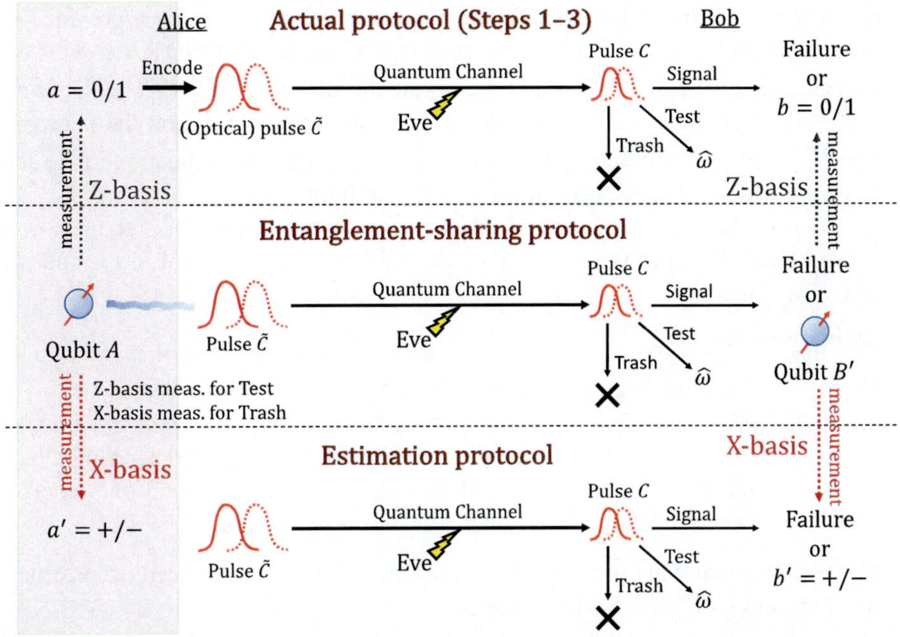

Fig. 4.6 Relation between the three protocols. The actual protocol and the estimation protocol are related through the entanglement-sharing protocol. After the entanglement-sharing protocol, Alice and Bob are left with the observed data ($\check{N}^{\rm suc}$, $\check{N}^{\rm fail}$, $\check{N}^{\rm test}$, $\check{N}^{\rm trash}$, $\check{F}$, $\check{Q}_-$) and $\check{N}^{\rm suc}$ pairs of qubits. If Alice and Bob ignore $\check{Q}_-$ and measure their qubits on the Z basis to determine their $\check{N}^{\rm suc}$-bit sifted keys, it becomes equivalent to the actual protocol. On the other hand, if Alice and Bob measure their $\check{N}^{\rm suc}$ pairs of qubits on the X basis, they can count the number $\check{N}^{\rm suc}_{\rm ph}$ of phase errors, which we call the estimation protocol. If we can find a reliable upper bound U on $\check{N}^{\rm suc}_{\rm ph}$ in the estimation protocol, it restricts the property of the state of $\check{N}^{\rm suc}$ pairs of qubits after the entanglement-sharing protocol, which in turn limits the amount of leaked information on the sifted keys in the actual protocol. The security proof is thus reduced to finding such an upper bound U in the estimation protocol, represented as a function of the variables that are commonly available in the three protocols

Alice's qubit, the resulting $\check{N}^{\rm fin}$-bit final key in this scenario is equivalent to that in the actual protocol. Although CNOT prevents Bob from obtaining an equivalent final key, he can still simulate the reconciliation and the verification process in Step 4 since the Z-basis value of each of his $\check{N}^{\rm suc}$ qubits corresponds to absence/presence of a bit error between Alice's and Bob's sifted key bits. Hence, Bob can equivalently carry out all the announcements in Steps 4 and 5 of the actual protocol. As a result, this scenario leads to exactly the same distribution $\Pr(\check{N}^{\rm fin} = N)$ and the same states $\rho^{\rm fin}_{AE|N}$ as those of the actual protocol, and thus corresponds to a virtual protocol in the sense of Sect. 4.3.4; i.e., Eq. (4.78) is satisfied with $\rho^{\rm virt}_{AE|N}$ being the state of Alice's qubits and Eve's system just before Alice's Z-basis measurement.

The secrecy condition (4.79) of Alice's final key can now be determined from the X-basis property of her $\check{N}^{\rm suc} = N'$ qubits after the application of CNOT, whose state is denoted by $\rho_{A|N'}$. Since CNOT can be rewritten as CNOT $= I_A \otimes |+\rangle\langle+|_B +$

$\sigma_A^Z \otimes |-\rangle\langle-|_B$, the X-basis values of $\rho_{A|N'}$ correspond to absence/presence of a phase error $\sigma_A^X \otimes \sigma_B^X = \pm 1$ just before the application of CNOT. Let $\check{\boldsymbol{x}}_A$ be the sequence obtained by the X-basis measurement on $\rho_{A|N'}$, and $\check{N}_{\mathrm{ph}}^{\mathrm{suc}}$ be the number of '$-$' symbols in $\check{\boldsymbol{x}}_A$. If we can have a good upper bound $\check{e}_{\mathrm{ph}}$ on the phase error rate $\check{N}_{\mathrm{ph}}^{\mathrm{suc}}/\check{N}^{\mathrm{suc}}$, shortening by fraction $h(\check{e}_{\mathrm{ph}})$ via privacy amplification in the actual protocol achieves the security in the asymptotic limit [73].

To cover the finite-size case as well, we need a more rigorous statement on the upper bound. For that purpose, we define an estimation protocol in the following (see also[5] Fig. 4.6).

Estimation protocol

$1'' - 3''$. Same as Steps $1'$, $2'$, and $3'$ of the entanglement-sharing protocol.

$4''$. Alice and Bob measure each of their $\check{N}^{\mathrm{suc}}$ pairs of qubits on the X basis and obtain outcomes a' and b', respectively. Let $\check{\boldsymbol{x}}_A$ be the binary sequence with each bit value being zero if $a' = b'$ and one otherwise. Define $\check{N}_{\mathrm{ph}}^{\mathrm{suc}} := \mathrm{wt}(\check{\boldsymbol{x}}_A)$, where $\mathrm{wt}(\cdot)$ is defined in Eq. (4.4).

The task of proving the security of the actual protocol is then reduced to construction of a function $U(\check{F}, \check{N}^{\mathrm{trash}})$ that satisfies

$$\Pr\left[\check{N}_{\mathrm{ph}}^{\mathrm{suc}} \leq U(\check{F}, \check{N}^{\mathrm{trash}})\right] \geq 1 - \epsilon \tag{4.122}$$

for any attack in the estimation protocol. In fact, this condition implies

$$\Pr\left[\check{N}^{\mathrm{suc}} \geq 1, \check{\boldsymbol{x}}_A \notin \mathcal{T}(\check{N}^{\mathrm{suc}}, \check{F}, \check{N}^{\mathrm{trash}})\right] \leq \epsilon, \tag{4.123}$$

where $\mathcal{T}(\check{N}^{\mathrm{suc}}, \check{F}, \check{N}^{\mathrm{trash}})$ denotes the set of all the possible patterns with $\mathrm{wt}(\check{\boldsymbol{x}}_A) \leq U(\check{F}, \check{N}^{\mathrm{trash}})$. Furthermore, from Lemma 4.2.3, it satisfies

$$\log|\mathcal{T}(\check{N}^{\mathrm{suc}}, \check{F}, \check{N}^{\mathrm{trash}})| \leq \check{N}^{\mathrm{suc}} h(\check{p}) \leq \lceil \check{N}^{\mathrm{suc}} h(\check{p}) \rceil, \tag{4.124}$$

with

$$\check{p} := \min\left\{U(\check{F}, \check{N}^{\mathrm{trash}})/\check{N}^{\mathrm{suc}}, 1/2\right\}. \tag{4.125}$$

Therefore, from Eqs. (4.96), (4.97), and (4.98), setting the final key length to

$$\check{N}^{\mathrm{fin}} = \max\{\check{N}^{\mathrm{suc}} - \lceil \check{N}^{\mathrm{suc}} h(\check{p}) \rceil - s, 0\} \tag{4.126}$$

achieves the $\varepsilon_{\mathrm{sec}} = \sqrt{2(\epsilon + 2^{-s})}$-secrecy.

To construct the function $U(\check{F}, \check{N}^{\mathrm{trash}})$ that satisfies Eq. (4.122), it is beneficial to clarify what property of the optical pulse C is measured by Bob's X-basis measurement in the estimation protocol (see Fig. 4.5). Let $\Pi_{\mathrm{ev(od)}}$ be the projection to the

[5] Modified from [46], Copyright 2021, The Author(s) licensed under CC BY 4.0.

subspace with even (resp. odd) photon numbers. ($\Pi_{\rm ev} + \Pi_{\rm od} = I_C$ holds by definition.) Furthermore, since $\Pi_{\rm ev} - \Pi_{\rm odd}$ is the operator for an optical phase shift of π, we have $\langle q| (\Pi_{\rm ev} - \Pi_{\rm odd}) = \langle -q|$. Eq. (4.119) is then rewritten as

$$K^{(q)} = \sqrt{2 f_{\rm suc}(q)} \left(|+\rangle_B \langle q|_C \Pi_{\rm ev} + |-\rangle_B \langle q|_C \Pi_{\rm od} \right). \tag{4.127}$$

Therefore, when the state of the pulse C is ρ_C, the probability of obtaining $+(-)$ in the X-basis measurement in the estimation protocol is given by

$$\langle +(-)| \mathcal{F}_{C\to B}(\rho_C) |+(-)\rangle = \mathrm{Tr}\left(\rho_C M^{\rm suc}_{\rm ev(od)}\right), \tag{4.128}$$

where

$$M^{\rm suc}_{\rm ev(od)} := \int_{-\infty}^{\infty} 2 f_{\rm suc}(q) dq \, \Pi_{\rm ev(od)} |q\rangle\langle q|_C \Pi_{\rm ev(od)}. \tag{4.129}$$

This shows that Bob's X-basis measurement distinguishes the parity of the photon number of the received pulse. In this sense, the secrecy of our protocol is assured by the complementarity between the sign of the quadrature and the parity of the photon number.

As an intermediate step toward our final goal of Eq. (4.122), let us first derive an upper bound on the expectation value $\mathbb{E}[\check{N}^{\rm suc}_{\rm ph}]$ in terms of those collected in the test and the trash rounds, $\mathbb{E}[\check{F}]$ and $\mathbb{E}[\check{Q}_-]$, in the estimation protocol. Let ρ_{AC} be the state of the qubit A and the received pulse C averaged over N pairs, and define relevant operators as

$$M^{\rm suc}_{\rm ph} := |+\rangle\langle +|_A \otimes M^{\rm suc}_{\rm od} + |-\rangle\langle -|_A \otimes M^{\rm suc}_{\rm ev}, \tag{4.130}$$

$$\Pi^{\rm fid} := |0\rangle\langle 0|_A \otimes |\beta\rangle\langle\beta|_C + |1\rangle\langle 1|_A \otimes |-\beta\rangle\langle -\beta|_C, \tag{4.131}$$

$$\Pi^{\rm trash}_- := |-\rangle\langle -|_A \otimes I_C. \tag{4.132}$$

Then we immediately have

$$\mathbb{E}_\rho[\check{N}^{\rm suc}_{\rm ph}] = p_{\rm sig} N \, \mathrm{Tr}\Big(\rho_{AC} M^{\rm suc}_{\rm ph}\Big) \tag{4.133}$$

and

$$\mathbb{E}_\rho[\check{Q}_-] = p_{\rm trash} N \, \mathrm{Tr}\big(\rho_{AC} \Pi^{\rm trash}_-\big), \tag{4.134}$$

while application of the property of Eq. (4.111) leads to

$$\mathbb{E}_\rho[\check{F}] \le p_{\rm test} N \, \mathrm{Tr}\big(\rho_{AC} \Pi^{\rm fid}\big). \tag{4.135}$$

Let us denote $\mathrm{Tr}\,(\rho_{AC} M)$ simply by $\langle M\rangle$ for any operator M. The set of points $\left(\langle M^{\rm suc}_{\rm ph}\rangle, \langle \Pi^{\rm fid}\rangle, \langle \Pi^{\rm trash}_-\rangle\right)$ for all the density operators ρ_{AC} form a convex region.

Rather than directly deriving the boundary of the region, it is easier to pursue linear constraints in the form of

$$\langle M_{\mathrm{ph}}^{\mathrm{suc}} \rangle \leq B(\kappa, \gamma) - \kappa \langle \Pi^{\mathrm{fid}} \rangle + \gamma \langle \Pi_{-}^{\mathrm{trash}} \rangle , \tag{4.136}$$

where $B(\kappa, \gamma), \kappa, \gamma \in \mathbb{R}$.

It is expected that a meaningful bound is obtained only for $\kappa, \gamma \geq 0$. Decreasing the fidelity $\langle \Pi^{\mathrm{fid}} \rangle$ should allow more room for eavesdropping, leading to a larger value of phase error rate $\langle M_{\mathrm{ph}}^{\mathrm{suc}} \rangle$. Hence Eq. (4.136) will give a good bound only when $\kappa \geq 0$. As for $\langle \Pi_{-}^{\mathrm{trash}} \rangle$, it only depends on the marginal state of Alice's qubit A, which is independent of the adversary's attack. We thus have

$$\langle \Pi_{-}^{\mathrm{trash}} \rangle = q_{-} := \| \langle - |_A | \Psi \rangle_{A\tilde{C}} \|^2 = (1 - e^{-2\mu})/2. \tag{4.137}$$

Since Alice's use of a stronger pulse leads to more leaked information, $\gamma \geq 0$ should be chosen for a good bound.

To find a function $B(\kappa, \gamma)$ satisfying Eq. (4.136), let us define an operator

$$M[\kappa, \gamma] := M_{\mathrm{ph}}^{\mathrm{suc}} + \kappa \Pi^{\mathrm{fid}} - \gamma \Pi_{-}^{\mathrm{trash}}. \tag{4.138}$$

Then Eq. (4.136) is rewritten as

$$\mathrm{Tr}\,(\rho_{AC} M[\kappa, \gamma]) \leq B(\kappa, \gamma). \tag{4.139}$$

This condition holds for any ρ_{AC} iff $M[\kappa, \gamma]$ satisfies an operator inequality

$$M[\kappa, \gamma] \leq B(\kappa, \gamma) I_{AC}. \tag{4.140}$$

Since $M[\kappa, \gamma]$ is a bounded operator, such $B(\kappa, \gamma)$ always exists. If the operator $M[\kappa, \gamma]$ were represented by a matrix of small size, the tightest bound would be found by computing the largest eigenvalue of the matrix. But here, $M[\kappa, \gamma]$ may not be a finite-rank operator (not even be a compact operator), and therefore it is difficult to compute the tightest bound. We thus compromise and heuristically find a computable bound $B(\kappa, \gamma)$ which is not necessarily tight; we reduce the problem to finding the largest eigenvalues of small-size matrices by replacing $M[\kappa, \gamma]$ with a constant upper bound except in a relevant finite-dimensional subspace spanned by $|\pm\beta\rangle$ and $M_{\mathrm{ev(od)}}^{\mathrm{suc}} |\pm\beta\rangle$. For the detailed derivation of $B(\kappa, \gamma)$, see the next section, Sect. 4.4.4.

With $B(\kappa, \gamma)$ computed, we can rewrite Eq. (4.136) using Eqs. (4.133)–(4.135) to obtain a relation between $\mathbb{E}[\check{N}_{\mathrm{ph}}^{\mathrm{suc}}]$, $\mathbb{E}[\check{F}]$, and $\mathbb{E}[\check{Q}_{-}]$. It is concisely written as

$$\mathbb{E}\big[\check{T}[\kappa, \gamma]\big] \leq N B(\kappa, \gamma) \tag{4.141}$$

with

$$\check{T}[\kappa, \gamma] := p_{\rm sig}^{-1}\check{N}_{\rm ph}^{\rm suc} + p_{\rm test}^{-1}\kappa\check{F} - p_{\rm trash}^{-1}\gamma\check{Q}_{-}. \tag{4.142}$$

This relation leads to an explicit bound on the phase error rate as $\mathbb{E}[\check{N}_{\rm ph}^{\rm suc}]/p_{\rm sig}N \leq B(\kappa, \gamma) + \gamma q_{-} - \kappa\,\mathbb{E}[\check{F}]/p_{\rm test}N$, which is enough for the computation of asymptotic key rates.

The security in the finite-size regime is proved by evaluating the fluctuations of $\check{T}[\kappa, \gamma]$ around the expectation value as follows. In the estimation protocol, the following random variables labeled by the number i of the round can be defined;

(i) $\check{N}_{\rm ph}^{{\rm suc},(i)}$ is defined to be unity only when "signal" is chosen in the i-th round, the detection is a "success", and a pair of outcomes (a', b') is $(+, -)$ or $(-, +)$. Otherwise, $\check{N}_{\rm ph}^{{\rm suc},(i)} = 0$. We have

$$\check{N}_{\rm ph}^{{\rm suc},(i)} = \begin{cases} 1 & (\text{signal, success, } (+,-) \text{ or } (-,+)) \\ 0 & (\text{otherwise}) \end{cases}, \tag{4.143}$$

and $\check{N}_{\rm ph}^{\rm suc} = \sum_{i=1}^{N}\check{N}_{\rm ph}^{{\rm suc},(i)}$.

(ii) $\check{F}^{(i)}$ is defined to be $\Lambda_{m,r}(|\check{\alpha} - (-1)^a\beta|^2)$ only when "test" is chosen in the i-th round. We have

$$\check{F}^{(i)} = \begin{cases} \Lambda_{m,r}(|\check{\alpha} - (-1)^a\beta|^2) & (\text{test}) \\ 0 & (\text{otherwise}) \end{cases}, \tag{4.144}$$

and $\check{F} = \sum_{i=1}^{N}\check{F}^{(i)}$.

(iii) $\check{Q}_{-}^{(i)}$ is defined to be unity only when "trash" is chosen in the i-th round and $a' = -$. Otherwise, $\check{Q}_{-}^{(i)} = 0$. We have

$$\check{Q}_{-}^{(i)} = \begin{cases} 1 & (\text{trash, } -) \\ 0 & (\text{otherwise}) \end{cases}, \tag{4.145}$$

and $\check{Q}_{-} = \sum_{i=1}^{N}\check{Q}_{-}^{(i)}$.

(iv) We also define

$$\check{T}^{(i)} := p_{\rm sig}^{-1}\check{N}_{\rm ph}^{{\rm suc},(i)} + p_{\rm test}^{-1}\kappa\check{F}^{(i)} - p_{\rm trash}^{-1}\gamma\check{Q}_{-}^{(i)}, \tag{4.146}$$

which leads to $\check{T}[\kappa, \gamma] = \sum_{i=1}^{N}\check{T}^{(i)}$.

Let $\sigma(\boldsymbol{Y})$ be a σ-algebra generated by a random vector $\boldsymbol{Y}$ (composed of n random variables), i.e.,

$$\sigma(\boldsymbol{Y}) := \{\boldsymbol{Y}^{-1}(A) \mid A \in \mathcal{B}(\mathbb{R}^n)\}. \tag{4.147}$$

Let

$$\{\varnothing, \Omega\} \subseteq \sigma(\check{T}^{(\leq 1)}) \subseteq \cdots \subseteq \sigma(\check{T}^{(\leq N)}) \tag{4.148}$$

be a filtration, where $\check{T}^{(\leq i)} := \left(\check{T}^{(1)}, \ldots, \check{T}^{(i)}\right)$ are random vectors. Then, we can apply Corollary 4.2.19 to the sequence $\left(\check{T}^{(1)}, \ldots, \check{T}^{(N)}\right)$ with respect to the filtration (4.148) if we can find constants $c_{\min}$ and $c_{\max}$ that satisfy

$$c_{\min} \leq \check{T}^{(i)} \leq c_{\max} \tag{4.149}$$

for any $i \in \{1, \ldots, N\}$. We define $c_{\min}$ and $c_{\max}$ as follows. In each round, at most one of $\check{N}_{\mathrm{ph}}^{\mathrm{suc},(i)}$, $\check{F}^{(i)}$, and $\check{Q}_{-}^{(i)}$ takes non-zero value; $\check{N}_{\mathrm{ph}}^{\mathrm{suc},(i)}$ and $\check{Q}_{-}^{(i)}$ are either zero or unity, and $\inf \Lambda_{m,r} \leq \check{F}^{(i)} \leq \sup \Lambda_{m,r}$. Since $\kappa, \gamma \geq 0$, Eq. (4.149) holds when $c_{\min}$ and $c_{\max}$ are defined as

$$c_{\min} := \min\{p_{\mathrm{test}}^{-1}\kappa \inf \Lambda_{m,r},\ -p_{\mathrm{trash}}^{-1}\gamma\}, \tag{4.150}$$

$$c_{\max} := \max\{p_{\mathrm{sig}}^{-1},\ p_{\mathrm{test}}^{-1}\kappa \sup \Lambda_{m,r}\}. \tag{4.151}$$

Applying Corollary 4.2.19 to the sequence $\left(\check{T}^{(1)}, \ldots, \check{T}^{(N)}\right)$ with respect to the filtration (4.148), we have, with a probability no smaller than $1 - \epsilon/2$,

$$\check{T}[\kappa, \gamma] \leq \sum_{i=1}^{N} \mathbb{E}[\check{T}^{(i)} | \sigma(\check{T}^{(\leq i-1)})] + \delta_1(\epsilon/2), \tag{4.152}$$

where

$$\delta_1(\epsilon) := (c_{\max} - c_{\min})\sqrt{\frac{N}{2} \ln\left(\frac{1}{\epsilon}\right)}. \tag{4.153}$$

Furthermore, for any state ρ_{AC}, we have

$$\mathbb{E}_\rho[\check{N}_{\mathrm{ph}}^{\mathrm{suc},(i)}] = p_{\mathrm{sig}} \mathrm{Tr}\left(\rho_{AC} M_{\mathrm{ph}}^{\mathrm{suc}}\right), \tag{4.154}$$

$$\mathbb{E}_\rho[\check{Q}_{-}^{(i)}] = p_{\mathrm{trash}} \mathrm{Tr}\left(\rho_{AC} \Pi_{-}^{\mathrm{trash}}\right), \tag{4.155}$$

$$\mathbb{E}_\rho[\check{F}^{(i)}] \leq p_{\mathrm{test}} \mathrm{Tr}\left(\rho_{AC} \Pi^{\mathrm{fid}}\right), \tag{4.156}$$

and thus

$$\mathbb{E}_\rho[\check{T}^{(i)}] \leq \mathrm{Tr}(\rho_{AC} M[\kappa, \gamma]) \leq B(\kappa, \gamma), \tag{4.157}$$

for $i \in \{1, \ldots, N\}$. Since the above inequality holds for any state ρ_{AC}, the following holds for each $i \in \{1, \ldots, N\}$ and for any $D \in \sigma(\check{T}^{(\leq i-1)})$:

$$\int_D \mathbb{E}[\check{T}^{(i)}|\sigma(\check{T}^{(\le i-1)})](x)\, P^{(\le i-1)}(dx) \le \mathrm{Tr}\left[\left(E^{(\le i-1)}(D) \otimes M[\kappa,\gamma] \otimes I^{\otimes(N-i)}\right)\rho^{\mathrm{tot}}\right] \tag{4.158}$$

$$\le B(\kappa,\gamma)\int_D P^{(\le i-1)}(dx), \tag{4.159}$$

where $E^{(\le i-1)}$ denotes the POVM for the random variables $\check{T}^{(\le i-1)}$, ρ^{tot} denotes the density operator for the total round with which the probability measure for Azuma's inequality is defined, and the probability measure $P^{(\le i-1)}(dx)$ is defined as

$$P^{(\le i-1)}(dx) := \mathrm{Tr}\left[\left(E^{(\le i-1)}(dx) \otimes I^{\otimes(N-i+1)}\right)\rho^{\mathrm{tot}}\right]. \tag{4.160}$$

The inequality (4.159) implies that the following inequality holds almost surely:

$$\mathbb{E}[\check{T}^{(i)}|\sigma(\check{T}^{(\le i-1)})] \le B(\kappa,\gamma). \tag{4.161}$$

Combining this with Eq. (4.152), we have

$$\check{T}[\kappa,\gamma] \le NB(\kappa,\gamma) + \delta_1(\epsilon/2), \tag{4.162}$$

which holds with a probability no smaller than $1-\epsilon/2$. We remark that the reason for including the trash rounds in the actual protocol is to circumvent a technical issue which would arise in this step. Without measurement of $\check{Q}_-$ in the estimation protocol, we would obtain an inequality $\mathbb{E}[p_{\mathrm{sig}}^{-1}\check{N}_{\mathrm{ph}}^{\mathrm{suc}} + p_{\mathrm{test}}^{-1}\kappa\check{F}] \le NB(\kappa,\gamma) + \gamma q_-$. In contrast to Eq. (4.141), the new inequality is true only when ρ_{AC} satisfies $\langle \Pi_-^{\mathrm{trash}}\rangle = q_-$, which is too stringent for the application of Azuma's inequality.

Although Eq. (4.142) includes $\check{Q}_-$ which is inaccessible in the actual protocol, we can derive an upper bound by noticing that it is an outcome from Alice's qubits and is independent of the adversary's attack. In fact, given $\check{N}^{\mathrm{trash}} = n$, it is the tally of n Bernoulli trials with a probability q_-, i.e., $\Pr(\check{Q}_- \mid \check{N}^{\mathrm{trash}} = n)$ is a binomial distribution. Thus from Corollary 4.2.10, the following inequality holds with a probability no smaller than $1-\epsilon/2$:

$$\check{Q}_- \le \check{N}^{\mathrm{trash}} q_- + \delta_2(\epsilon/2;\check{N}^{\mathrm{trash}}), \tag{4.163}$$

where

$$\delta_2(\epsilon/2;\check{N}^{\mathrm{trash}}) := \check{N}^{\mathrm{trash}}\delta(\epsilon/2;\check{N}^{\mathrm{trash}}) \tag{4.164}$$

with $\delta(\epsilon,n)$ defined in Corollary 4.2.10. Note that when $\check{N}^{\mathrm{trash}}$ is equal to zero, $\check{Q}_-$ is also equal to zero, and thus Eq. (4.163) also holds.

Combining Eqs. (4.142), (4.162), and (4.163), we obtain $U(\check{F},\check{N}^{\mathrm{trash}})$, satisfying Eq. (4.122) by the union bound, as follows:

$$U(\check{F}, \check{N}^{\mathrm{trash}}) = p_{\mathrm{sig}}(NB(\kappa, \gamma) + \delta_1(\epsilon/2)) - \frac{p_{\mathrm{sig}}}{p_{\mathrm{test}}}\kappa \check{F} + \frac{p_{\mathrm{sig}}}{p_{\mathrm{trash}}}\gamma\left(\check{N}^{\mathrm{trash}}q_- + \delta_2(\epsilon/2; \check{N}^{\mathrm{trash}})\right). \tag{4.165}$$

We finally remark that the encryption of $M := N_{\mathrm{EC}} + s'$ bits in Step 4 can be omitted as long as each bit linearly depends on Alice's sifted key over GF(2). In such a case, the virtual and the estimation protocol must include measurements on Alice's qubits to simulate the announcement of the M bits in Step 4. (Otherwise, the condition (4.78) cannot be met.) The back-action on the X basis caused by the measurement for each bit amounts to doubling the number of probable patterns $\check{\boldsymbol{x}}_A$. We can thus redefine the set $\mathcal{T}(\check{N}^{\mathrm{suc}}, \check{F}, \check{N}^{\mathrm{trash}})$ by enlarging its size by factor of 2^M so that Eq. (4.123) holds. This leads to adding M to $N_{\mathrm{PA}} := \lceil \check{N}^{\mathrm{suc}} h(p) \rceil$, which eventually reduces the length of $\check{N}^{\mathrm{fin}}$ by M-bits. This means that we achieve the same net key gain rate $\check{G}$ with the same level of security.

4.4.4 Derivation of the Operator Inequality

The goal of this section is to explicitly construct $B(\kappa, \gamma)$ satisfying (4.140), which was the heuristic part of the security proof in the previous section. Let $\lambda_{\mathrm{sup}}(O)$ denote the supremum of the spectrum of a bounded self-adjoint operator O. Although $B(\kappa, \gamma) = \lambda_{\mathrm{sup}}(M[\kappa, \gamma])$ would give the tightest bound satisfying Eq. (4.140), it is hard to compute it numerically since system C has an infinite-dimensional Hilbert space. Instead, we derive a looser but simpler bound. We first prove the following lemma.

Lemma 4.4.3 *Let $\Pi_\pm$ be orthogonal projections satisfying $\Pi_+\Pi_- = 0$. Suppose that the rank of $\Pi_\pm$ is no smaller than two or infinite. Let $M_\pm$ be self-adjoint operators satisfying $\Pi_\pm M_\pm \Pi_\pm = M_\pm \le \alpha_\pm \Pi_\pm$, where $\alpha_\pm$ are real constants. Let $|\psi\rangle$ be an unnormalized vector satisfying $(\Pi_+ + \Pi_-)|\psi\rangle = |\psi\rangle$ and $\Pi_\pm |\psi\rangle \neq 0$. Define the following quantities with respect to $|\psi\rangle$:*

$$C_\pm := \langle\psi| \Pi_\pm |\psi\rangle \ (> 0), \tag{4.166}$$

$$D_\pm := C_\pm^{-1} \langle\psi| M_\pm |\psi\rangle, \tag{4.167}$$

$$V_\pm := C_\pm^{-1} \langle\psi| M_\pm^2 |\psi\rangle - D_\pm^2. \tag{4.168}$$

Then, for any real numbers γ_+ and γ_-, we have

$$\lambda_{\mathrm{sup}}\big(M_+ + M_- + |\psi\rangle\langle\psi| - \gamma_+\Pi_+ - \gamma_-\Pi_-\big) \le \lambda_{\mathrm{sup}}\big(M_{4\mathrm{d}}\big), \tag{4.169}$$

where four dimensional matrix $M_{4\mathrm{d}}$ is defined as

$$M_{4\mathrm{d}} := \begin{pmatrix} \alpha_+ - \gamma_+ & \sqrt{V_+} & 0 & 0 \\ \sqrt{V_+} & C_+ + D_+ - \gamma_+ & \sqrt{C_+C_-} & 0 \\ 0 & \sqrt{C_+C_-} & C_- + D_- - \gamma_- & \sqrt{V_-} \\ 0 & 0 & \sqrt{V_-} & \alpha_- - \gamma_- \end{pmatrix}. \tag{4.170}$$

Proof We choose orthonormal vectors $\{|e_\pm^{(1)}\rangle, |e_\pm^{(2)}\rangle\}$ in the domain of $\Pi_\pm$, respectively, to satisfy

$$\sqrt{C_\pm}\,|e_\pm^{(1)}\rangle = \Pi_\pm\,|\psi\rangle, \tag{4.171}$$

$$M_\pm\,|e_\pm^{(1)}\rangle = D_\pm\,|e_\pm^{(1)}\rangle + \sqrt{V_\pm}\,|e_\pm^{(2)}\rangle, \tag{4.172}$$

which is well-defined due to Eqs. (4.166)–(4.168) and $\Pi_\pm M_\pm \Pi_\pm = M_\pm$. From $(\Pi_+ + \Pi_-)\,|\psi\rangle = |\psi\rangle$, we have

$$|\psi\rangle = \sqrt{C_+}\,|e_+^{(1)}\rangle + \sqrt{C_-}\,|e_-^{(1)}\rangle. \tag{4.173}$$

Let us define the following projection operators:

$$\Pi_\pm^{(j)} := |e_\pm^{(j)}\rangle\langle e_\pm^{(j)}| \quad (j = 1, 2), \tag{4.174}$$

$$\Pi_\pm^{(\geq 2)} := \Pi_\pm - \Pi_\pm^{(1)}, \tag{4.175}$$

$$\Pi_\pm^{(\geq 3)} := \Pi_\pm^{(\geq 2)} - \Pi_\pm^{(2)}. \tag{4.176}$$

Since Eq. (4.172) implies $\Pi_\pm^{(\geq 3)} M_\pm \Pi_\pm^{(1)} = 0$, we have

$$M_\pm = \Pi_\pm^{(1)} M_\pm \Pi_\pm^{(1)} + \Pi_\pm^{(\geq 2)} M_\pm \Pi_\pm^{(\geq 2)} + \Pi_\pm^{(1)} M_\pm \Pi_\pm^{(2)} + \Pi_\pm^{(2)} M_\pm \Pi_\pm^{(1)}. \tag{4.177}$$

The second term in the right-hand side of Eq. (4.177) is bounded as

$$\Pi_\pm^{(\geq 2)} M_\pm \Pi_\pm^{(\geq 2)} \leq \alpha_\pm \Pi_\pm^{(\geq 2)}, \tag{4.178}$$

since $M_\pm \leq \alpha_\pm \Pi_\pm$. Combining Eqs. (4.167), (4.177), and (4.178), we have

$$\begin{aligned} M_\pm - \gamma_\pm \Pi_\pm \leq{} & (D_\pm - \gamma_\pm)\,|e_\pm^{(1)}\rangle\langle e_\pm^{(1)}| + (\alpha_\pm - \gamma_\pm)\,|e_\pm^{(2)}\rangle\langle e_\pm^{(2)}| \\ & + \sqrt{V_\pm}\big(|e_\pm^{(1)}\rangle\langle e_\pm^{(2)}| + |e_\pm^{(2)}\rangle\langle e_\pm^{(1)}|\big) + (\alpha_\pm - \gamma_\pm)\Pi_\pm^{(\geq 3)}. \end{aligned} \tag{4.179}$$

Combining Eqs. (4.173) and (4.179), we have

$$\begin{aligned} & M_+ + M_- + |\psi\rangle\langle\psi| - \gamma_+\Pi_+ - \gamma_-\Pi_- \\ & \leq M_{4\mathrm{d}} \oplus (\alpha_+ - \gamma_+)\Pi_+^{(\geq 3)} \oplus (\alpha_- - \gamma_-)\Pi_-^{(\geq 3)}, \end{aligned} \tag{4.180}$$

where M_{4d} is given in Eq. (4.170) with the basis $\left\{|e_+^{(2)}\rangle, |e_+^{(1)}\rangle, |e_-^{(1)}\rangle, |e_-^{(2)}\rangle\right\}$. Since $\alpha_\pm - \gamma_\pm = \langle e_\pm^{(2)}| M_{4d} |e_\pm^{(2)}\rangle \leq \lambda_{\mathrm{sup}}(M_{4d})$, supremum of the spectrum of the right-hand side of Eq. (4.180) is equal to the largest eigenvalue of the four-dimensional matrix M_{4d}. We thus obtain Eq. (4.169). □

As a corollary, we derive Eq. (4.140) as follows.

Corollary 4.4.4 *Let $|\beta\rangle$ be a coherent state. Let $\Pi_{\mathrm{ev(od)}}$, $M^{\mathrm{suc}}_{\mathrm{ev(od)}}$, and $M[\kappa, \gamma]$ be as defined in the main text, and define following quantities:*

$$C_{\mathrm{ev}} := \langle\beta|\, \Pi_{\mathrm{ev}}\, |\beta\rangle = e^{-|\beta|^2} \cosh |\beta|^2, \tag{4.181}$$

$$C_{\mathrm{od}} := \langle\beta|\, \Pi_{\mathrm{od}}\, |\beta\rangle = e^{-|\beta|^2} \sinh |\beta|^2, \tag{4.182}$$

$$D_{\mathrm{ev(od)}} := C^{-1}_{\mathrm{ev(od)}} \langle\beta|\, M^{\mathrm{suc}}_{\mathrm{ev(od)}}\, |\beta\rangle\,, \tag{4.183}$$

$$V_{\mathrm{ev(od)}} := C^{-1}_{\mathrm{ev(od)}} \langle\beta|\, \left(M^{\mathrm{suc}}_{\mathrm{ev(od)}}\right)^2 |\beta\rangle - D^2_{\mathrm{ev(od)}}. \tag{4.184}$$

Let $M^{\mathrm{err}}_{4d}[\kappa, \gamma]$ and $M^{\mathrm{cor}}_{2d}[\kappa, \gamma]$ be defined as follows:

$$M^{\mathrm{err}}_{4d}[\kappa, \gamma] := \begin{pmatrix} 1 & \sqrt{V_{\mathrm{od}}} & & \\ \sqrt{V_{\mathrm{od}}} & \kappa\, C_{\mathrm{od}} + D_{\mathrm{od}} & \kappa\sqrt{C_{\mathrm{od}}\, C_{\mathrm{ev}}} & \\ & \kappa\sqrt{C_{\mathrm{od}}\, C_{\mathrm{ev}}} & \kappa\, C_{\mathrm{ev}} + D_{\mathrm{ev}} - \gamma & \sqrt{V_{\mathrm{ev}}} \\ & & \sqrt{V_{\mathrm{ev}}} & 1-\gamma \end{pmatrix}, \tag{4.185}$$

$$M^{\mathrm{cor}}_{2d}[\kappa, \gamma] := \begin{pmatrix} \kappa\, C_{\mathrm{ev}} & \kappa\sqrt{C_{\mathrm{ev}}\, C_{\mathrm{od}}} \\ \kappa\sqrt{C_{\mathrm{ev}}\, C_{\mathrm{od}}} & \kappa\, C_{\mathrm{od}} - \gamma \end{pmatrix}. \tag{4.186}$$

Define a convex function

$$B(\kappa, \gamma) := \max\left\{\lambda_{\mathrm{sup}}\left(M^{\mathrm{err}}_{4d}[\kappa, \gamma]\right), \lambda_{\mathrm{sup}}\left(M^{\mathrm{cor}}_{2d}[\kappa, \gamma]\right)\right\}. \tag{4.187}$$

Then, for $\kappa, \gamma \geq 0$, we have

$$M[\kappa, \gamma] \leq B(\kappa, \gamma) I_{AC}. \tag{4.188}$$

Proof Let us first observe that the operator Π^{fid} defined in Eq. (4.131) can be rewritten as follows:

$$\Pi^{\mathrm{fid}} = |\phi_{\mathrm{err}}\rangle\langle\phi_{\mathrm{err}}|_{AC} + |\phi_{\mathrm{cor}}\rangle\langle\phi_{\mathrm{cor}}|_{AC}\,, \tag{4.189}$$

where orthogonal vectors $|\phi_{\mathrm{err}}\rangle_{AC}$ and $|\phi_{\mathrm{cor}}\rangle_{AC}$ are defined as

$$|\phi_{\mathrm{err}}\rangle_{AC} := |+\rangle_A \otimes \Pi_{\mathrm{od}}\, |\beta\rangle_C + |-\rangle_A \otimes \Pi_{\mathrm{ev}}\, |\beta\rangle_C\,, \tag{4.190}$$

$$|\phi_{\mathrm{cor}}\rangle_{AC} := |+\rangle_A \otimes \Pi_{\mathrm{ev}}\, |\beta\rangle_C + |-\rangle_A \otimes \Pi_{\mathrm{od}}\, |\beta\rangle_C\,. \tag{4.191}$$

Next, using Eqs. (4.189), (4.130), and (4.132), we rearrange the operator $M[\kappa, \gamma]$ defined in Eq. (4.138) as follows:

$$M[\kappa, \gamma] = M^{\text{err}}[\kappa, \gamma] \oplus M^{\text{cor}}[\kappa, \gamma], \tag{4.192}$$

where

$$\begin{aligned} M^{\text{err}}[\kappa, \gamma] &:= |+\rangle\langle+|_A \otimes M^{\text{suc}}_{\text{od}} + |-\rangle\langle-|_A \otimes M^{\text{suc}}_{\text{ev}} \\ &\quad + \kappa\, |\phi_{\text{err}}\rangle\langle\phi_{\text{err}}|_{AC} - \gamma\, |-\rangle\langle-|_A \otimes \Pi_{\text{ev}}, \end{aligned} \tag{4.193}$$

$$M^{\text{cor}}[\kappa, \gamma] := \kappa\, |\phi_{\text{cor}}\rangle\langle\phi_{\text{cor}}|_{AC} - \gamma\, |-\rangle\langle-|_A \otimes \Pi_{\text{od}}. \tag{4.194}$$

We can apply Lemma 4.4.3 to $M^{\text{err}}[\kappa, \gamma]$ by the following substitutions

$$M_\pm = |\pm\rangle\langle\pm|_A \otimes M^{\text{suc}}_{\text{od(ev)}}, \tag{4.195}$$
$$|\psi\rangle = \sqrt{\kappa}\, |\phi_{\text{err}}\rangle_{AC}, \tag{4.196}$$
$$\Pi_\pm = |\pm\rangle\langle\pm|_A \otimes \Pi_{\text{od(ev)}}, \tag{4.197}$$
$$\alpha_\pm = 1, \tag{4.198}$$
$$\gamma^+ = 0, \quad \gamma^- = \gamma. \tag{4.199}$$

Here, $M_\pm \le \Pi_\pm$ (i.e., $\alpha_\pm = 1$) holds because $M^{\text{suc}}_{\text{od(ev)}}$ are POVM elements. The other prerequisites of Lemma 4.4.3 are easy to be confirmed. Thus, we obtain

$$\lambda_{\text{sup}}\big(M^{\text{err}}[\kappa, \gamma]\big) \le \lambda_{\text{sup}}\big(M^{\text{err}}_{\text{4d}}[\kappa, \gamma]\big). \tag{4.200}$$

In the same way, we can apply Lemma 4.4.3 to $M^{\text{cor}}[\kappa, \gamma]$ via

$$M_\pm = 0, \tag{4.201}$$
$$|\psi\rangle = \sqrt{\kappa}\, |\phi_{\text{cor}}\rangle_{AC}, \tag{4.202}$$
$$\Pi_\pm = |\pm\rangle\langle\pm|_A \otimes \Pi_{\text{ev(od)}}, \tag{4.203}$$
$$\alpha_\pm = 0, \tag{4.204}$$
$$\gamma^+ = 0, \quad \gamma^- = \gamma. \tag{4.205}$$

Since $M_\pm = 0$ implies $D_\pm = V_\pm = 0$ in Lemma 4.4.3, this time we can reduce the dimension of relevant matrix Eq. (4.170) by separating known eigenvalues 0 and $-\gamma$. Therefore, we have

$$\lambda_{\text{sup}}\big(M^{\text{cor}}[\kappa, \gamma]\big) \le \max\big\{\lambda_{\text{sup}}\big(M^{\text{cor}}_{\text{2d}}[\kappa, \gamma]\big), 0, -\gamma\big\} = \lambda_{\text{sup}}\big(M^{\text{cor}}_{\text{2d}}[\kappa, \gamma]\big), \tag{4.206}$$

where the last inequality holds since $\gamma \ge 0$ and $\kappa C_{\text{ev}} \ge 0$. We then obtain Eq. (4.188) from Eqs. (4.192), (4.200), and (4.206). Since $M^{\text{err}}_{\text{4d}}[\kappa, \gamma]$ and $M^{\text{cor}}_{\text{2d}}[\kappa, \gamma]$ are symmetric and their elements linearly depend on κ and γ, $\lambda_{\text{sup}}\big(M^{\text{err}}_{\text{4d}}[\kappa, \gamma]\big)$ and $\lambda_{\text{sup}}\big(M^{\text{cor}}_{\text{2d}}[\kappa, \gamma]\big)$ are convex functions over κ and γ, and so is $B(\kappa, \gamma)$. □

4.4.5 Numerical Simulations

We simulated the net key gain per pulse $\check{G}$ as a function of the transmission distance L in the optical channel. We assume a channel model with a loss with the transmissivity $\eta = 10^{-0.02L}$ (including the efficiency of Bob's apparatus) and an excess noise at channel output; Bob receives Gaussian states obtained by randomly displacing coherent states $|\pm\sqrt{\eta\mu}\rangle$ to increase their variances by a factor of $(1+\xi)$ [78, 79]. The states that Bob receives dependent on Alice's bit value a are thus given by

$$\rho_{\text{model}}^{(a)} := \int_{\mathbb{C}} p_\xi(\gamma)\, |(-1)^a\sqrt{\eta\mu}+\gamma\rangle\langle(-1)^a\sqrt{\eta\mu}+\gamma|\, d^2\gamma, \tag{4.207}$$

where $p_\xi(\gamma)$ is given by

$$p_\xi(\gamma) := \frac{1}{\pi\xi} e^{-|\gamma|^2/\xi}. \tag{4.208}$$

The parameter ξ is the excess noise relative to the vacuum, i.e.,

$$\langle(\Delta\hat{q})^2\rangle_{\rho_{\text{model}}^{(a)}} = (1+\xi)/2, \tag{4.209}$$

where $\langle(\Delta\hat{q})^2\rangle = 1/2$ for the vacuum state from Eq. (3.86). The expected amplitude of coherent state β is chosen to be $\sqrt{\eta\mu}$. The actual fidelity between Bob's objective state $|(-1)^a\sqrt{\eta\mu}\rangle$ and the model state $\rho_{\text{model}}^{(a)}$ is given by

$$\begin{aligned} & F(\rho_{\text{model}}^{(a)}, |(-1)^a\sqrt{\eta\mu}\rangle\langle(-1)^a\sqrt{\eta\mu}|) \\ & = \int_{\mathbb{C}} p_\xi(\gamma) |\, \langle(-1)^a\sqrt{\eta\mu}|(-1)^a\sqrt{\eta\mu}-\gamma\rangle\, |^2 d\gamma \\ & = \frac{1}{1+\xi/2}. \end{aligned} \tag{4.210}$$

We assume a step function with a threshold $q_{\text{th}}(>0)$ as the acceptance probability, i.e., $f_{\text{suc}}(q) = \Theta(q-q_{\text{th}})$. In this case, the quantities defined in Eqs. (4.183) and (4.184) are given by

$$D_{\text{ev}} = \int_{-\infty}^{\infty} 2C_{\text{ev}}^{-1} f_{\text{suc}}(q)\, \left|\langle q|\, \Pi_{\text{ev}}\, |\beta\rangle\right|^2 dq \tag{4.211}$$

$$= \frac{1}{4C_{\text{ev}}}\left[\text{erfc}\left(q_{\text{th}}-\sqrt{2}\beta\right) + \text{erfc}\left(q_{\text{th}}+\sqrt{2}\beta\right) + 2e^{-2\beta^2}\text{erfc}\left(q_{\text{th}}\right)\right], \tag{4.212}$$

$$D_{\text{od}} = \int_{-\infty}^{\infty} 2C_{\text{od}}^{-1} f_{\text{suc}}(q)\, \left|\langle q|\, \Pi_{\text{od}}\, |\beta\rangle\right|^2 dq \tag{4.213}$$

$$= \frac{1}{4C_{\text{od}}}\left[\operatorname{erfc}\left(q_{\text{th}} - \sqrt{2}\beta\right) + \operatorname{erfc}\left(q_{\text{th}} + \sqrt{2}\beta\right) - 2e^{-2\beta^2}\operatorname{erfc}\left(q_{\text{th}}\right)\right], \quad (4.214)$$

$$V_{\text{ev(od)}} = \int_{-\infty}^{\infty} 2C_{\text{ev(od)}}^{-1}\left(f_{\text{suc}}(q)\right)^2 \left|\langle q|\,\Pi_{\text{ev(od)}}\,|\beta\rangle\right|^2 dq - D_{\text{ev(od)}}^2 \quad (4.215)$$

$$= D_{\text{ev(od)}} - D_{\text{ev(od)}}^2, \quad (4.216)$$

where $\beta = \sqrt{\eta\mu}$ and the complementary error function $\operatorname{erfc}(x)$ is defined as

$$\operatorname{erfc}(x) := \frac{2}{\sqrt{\pi}}\int_{x}^{\infty} dt\, e^{-t^2}. \quad (4.217)$$

For the derivation of Eq. (4.215), we used the fact that $\Pi_{\text{ev}} + \Pi_{\text{od}} = I$ and $(\Pi_{\text{ev}} - \Pi_{\text{od}})\,|\beta\rangle = |-\beta\rangle$.

We assume that the number of "success" signal rounds $\check{N}^{\text{suc}}$ is equal to its expectation value,

$$\mathbb{E}[\check{N}^{\text{suc}}] = \left(\int_{-\infty}^{\infty} f(q)\,\langle q|\,\rho_{\text{model}}^{(a)}\,|q\rangle\, dq\right) p_{\text{sig}}N$$

$$= p_{\text{sig}}N(P^{+} + P^{-}), \quad (4.218)$$

where

$$P^{\pm} := \int_{q_{\text{th}}}^{\infty} \langle \pm(-1)^a q|\,\rho_{\text{model}}^{(a)}\,|\pm(-1)^a q\rangle\, dq \quad (4.219)$$

$$= \frac{1}{2}\operatorname{erfc}\left((q_{\text{th}} \mp \sqrt{2}\sqrt{\eta\mu})\sqrt{\frac{1}{1+\xi}}\right). \quad (4.220)$$

We also assume that the number of test rounds $\check{N}^{\text{test}}$ is equal to $p_{\text{test}}N$ and the number of trash rounds $\check{N}^{\text{trash}}$ is equal to $p_{\text{trash}}N$. The test outcome $\check{F}$ is assumed to be equal to its expectation value $\mathbb{E}[\check{F}]$, which is given by

$$\mathbb{E}[\check{F}] = p_{\text{test}}N\,\mathbb{E}_{\rho_{\text{model}}^{(a)}}[\Lambda_{m,r}(|\check{\alpha} - (-1)^a\sqrt{\eta\mu}|^2)]$$

$$= p_{\text{test}}N\int_{\mathbb{C}} \frac{d^2\alpha}{\pi}\,\langle\alpha|\,\rho_{\text{model}}^{(a)}\,|\alpha\rangle\,\Lambda_{m,r}(|\alpha - (-1)^a\sqrt{\eta\mu}|^2)$$

$$= \frac{p_{\text{test}}N}{1+\xi/2}\left[1 - (-1)^{m+1}\left(\frac{\xi/2}{1+r(1+\xi/2)}\right)^{m+1}\right]. \quad (4.221)$$

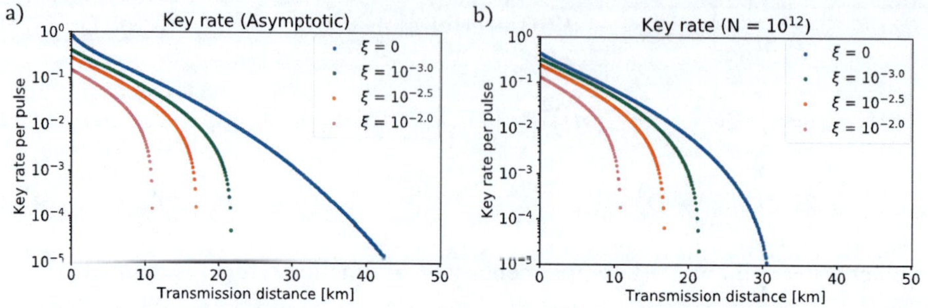

Fig. 4.7 The net key gain per pulse $\check{G}$ (key rate) vs. the transmission distance L of the optical channel. We assumed that the optical pulse that Bob receives is given by randomly displacing a coherent state to increase its variance by a factor of $(1+\xi)$. **a** The asymptotic key rate for various values of ξ. **b** The key rate for various values of ξ when the pulse number is finite ($N = 10^{12}$)

We adopted $m = 1$ and $r = 0.4120$ for $\Lambda_{m,r}$, which leads to $(\sup \Lambda_{m,r}, \inf \Lambda_{m,r}) = (2.824, -0.9932)$. The cost of the bit error correction $N_{\rm EC}$ is assumed to be $1.1 \times \check{N}^{\rm suc} h(e_{\rm bit})$, where the bit error rate $e_{\rm bit}$ is given by

$$e_{\rm bit} = \frac{P^-}{P^+ + P^-}. \tag{4.222}$$

We set $\varepsilon_{\rm sct} (:= \varepsilon_{\rm cor} + \varepsilon_{\rm sec}) = 2^{-50}$ for the overall security parameter, and set $\epsilon = 2^{-s} = \varepsilon_{\rm sec}^2/4 = \varepsilon_{\rm sct}^2/16$ and $2^{-s'} = \varepsilon_{\rm cor} = \varepsilon_{\rm sct}/2$. We thus have two coefficients (κ, γ) and four protocol parameters $(\mu, q_{\rm th}, p_{\rm sig}, p_{\rm test})$ to be determined. For each transmission distance L, we determined (κ, γ) via a convex optimization using the CVXPY 1.0.25 [80, 81] and $(\mu, q_{\rm th}, p_{\rm sig}, p_{\rm test})$ via the Nelder-Mead in the scipy.minimize library in Python, in order to maximize the key rate.

Figures 4.7[6] and 4.8[7] show the key rates of our protocol in the asymptotic limit $N \to \infty$ and finite-size cases with $N = 10^9$–10^{12} for $\xi = 10^{-2.0}$–$10^{-3.0}$ and 0. (Note that from the results of the recent experiments [15, 79, 82], excess noise with $\xi = 10^{-2.0}$–$10^{-3.0}$ at the channel output seems reasonable. Furthermore, the state-of-the-art experiments [15] work at 0.5 GHz repetition rate, which implies that total number of rounds $N = 10^9$–10^{12} can be achieved in a realistic duration.) For the noiseless model ($\xi = 0$) in Fig. 4.7a, the asymptotic rate reaches 8 dB. In the case of $\xi = 10^{-3.0}$, it reaches 4 dB, which is comparable to the result of a similar binary modulation protocol proposed in Ref. [38].

As for finite-size key rates, we see that the noiseless model shows a significant finite-size effect even for $N = 10^{12}$ in Fig. 4.8a. On the other hand, with the presence of noises ($\xi = 10^{-3.0}$), the effect becomes milder, and $N = 10^{11}$ is enough to achieve a rate close to the asymptotic case, as can be seen in Fig. 4.8b. This may be ascribed to the cost of the fidelity test. In order to make sure that the fidelity is no smaller

[6] Modified from [46], Copyright 2021, The Author(s) licensed under CC BY 4.0.

[7] Modified from [46], Copyright 2021, The Author(s) licensed under CC BY 4.0.

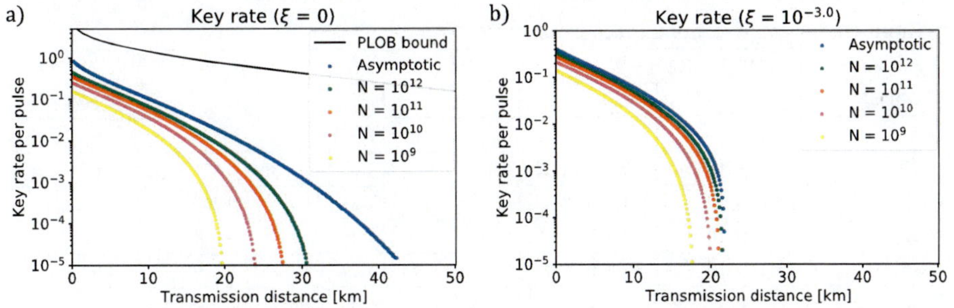

Fig. 4.8 The net key gain per pulse $\check{G}$ (key rate) vs. the transmission distance L of the optical channel. We assumed that the optical pulse that Bob receives is given by randomly displacing a coherent state to increase its variance by a factor of $(1+\xi)$. **a** The key rate in a pure-loss channel ($\xi = 0$) for various pulse numbers N. The solid black line shows the fundamental limitation of the one-way repeater-less secure key distribution called the PLOB bound [4]. **b** The key rate in a channel with the excess noise $\xi = 10^{-3.0}$ for various pulse numbers N

than $1-\delta$, the statistical uncertainty of the fidelity test must be reduced to $O(\delta)$. As a result, approaching the asymptotic rate of $\xi = 0$ will require many rounds for the fidelity test.

Examples of optimized parameters are shown in Table 4.1.[8] Typical optimized values of the threshold q_{th} range from 0.7 to 2.0. They are larger than those in other analyses of protocols with post-selection (e.g., [40]). A possible reason is that the latter protocols use more than two states to monitor the eavesdropping action, which may lead to a lower cost of privacy amplification and higher tolerance against bit errors.

4.4.6 Discussion

Numerically simulated key rates above were computed on the implicit assumption that Bob's observed quantities are processed with infinite precision. Even when these are approximated with a finite set of discrete points, we can still prove the security with minimal degradation of key rates. We have already explained in Sect. 4.4.1 the capability to incorporate finite precision for the heterodyne measurement used for the test in the protocol. For the homodyne measurement used for the signal, finite precision can be treated through appropriate modification of the acceptance probability $f_{\text{suc}}(q)$. Aside from a small change in the success rate and the bit error rate, this function affects the key rate only through integrals in Eqs. (4.211), (4.213), and (4.215) in the previous section, and hence influence on the key rate is expected to be small. We thus believe that the fundamental obstacles associated with the analog nature of the continuous-variable protocol have been settled by our approach.

[8] Modified from [46], Copyright 2021, The Author(s) licensed under CC BY 4.0.

Table 4.1 Examples of optimized parameters

Parameters for $N = 10^{11}$ and $\xi = 0$

L [km]	Key rate $\check{G}$	(κ, γ)	μ	q_{th}	p_{sig}	p_{test}
5	1.38×10^{-1}	(32.4, 1.38)	0.514	0.752	0.821	0.172
10	5.36×10^{-2}	(20.3, 0.741)	0.442	1.02	0.831	0.160
15	1.75×10^{-2}	(14.6, 0.381)	0.451	1.21	0.767	0.227
20	4.15×10^{-3}	(12.6, 0.175)	0.442	1.46	0.624	0.371
25	3.23×10^{-4}	(9.98, 0.059)	0.438	1.77	0.370	0.626

Parameters for $N = 10^{11}$ and $\xi = 10^{-3.0}$

L [km]	Key rate $\check{G}$	(κ, γ)	μ	q_{th}	p_{sig}	p_{test}
5	1.13×10^{-1}	(17.9, 1.19)	0.466	0.802	0.853	0.138
10	4.07×10^{-2}	(11.9, 0.666)	0.442	1.02	0.831	0.160
15	1.02×10^{-2}	(8.85, 0.345)	0.440	1.27	0.758	0.233
20	5.29×10^{-4}	(6.70, 0.147)	0.463	1.57	0.484	0.505

Parameters for $N = 10^{12}$ and $\xi = 0$

L [km]	Key rate $\check{G}$	(κ, γ)	μ	q_{th}	p_{sig}	p_{test}
5	1.69×10^{-1}	(54.9, 1.56)	0.556	0.713	0.874	0.123
10	6.76×10^{-2}	(31.3, 0.798)	0.493	0.948	0.868	0.128
15	2.41×10^{-2}	(23.2, 0.408)	0.466	1.18	0.829	0.168
20	6.93×10^{-3}	(17.7, 0.184)	0.443	1.43	0.772	0.226
25	1.14×10^{-3}	(14.6, 0.065)	0.427	1.71	0.596	0.402
30	3.23×10^{-5}	(10.8, 0.017)	0.421	2.05	0.240	0.759

Parameters for $N = 10^{12}$ and $\xi = 10^{-3.0}$

L [km]	Key rate $\check{G}$	(κ, γ)	μ	q_{th}	p_{sig}	p_{test}
5	1.32×10^{-1}	(21.7, 1.25)	0.482	0.783	0.909	0.086
10	4.92×10^{-2}	(13.8, 0.689)	0.450	1.01	0.899	0.096
15	1.36×10^{-2}	(9.82, 0.355)	0.442	1.25	0.858	0.137
20	1.17×10^{-3}	(7.13, 0.151)	0.458	1.56	0.701	0.293

Examples of the parameters for a given pair of the total pulse number N and the excess noise parameter ξ defined in Eq. (4.209) The variance of the quadrature operator $\hat{q}$ for the vacuum state is $\langle(\Delta\hat{q})^2\rangle = 1/2$. Given (N, ξ), protocol parameters $(\kappa, \gamma, \mu, q_{\text{th}}, p_{\text{sig}}, p_{\text{test}})$ are optimized for each transmission distance L [km] so that the net key gain per pulse (key rate) $\check{G}$ is maximized

In comparison with recent asymptotic analyses [40, 41] of continuous-variable QKD with the discrete modulation, our protocol achieves lower key rates and much shorter distance. Since ours is the first attempt to apply the proof technique of discrete-variable QKD to continuous-variable QKD, there is much room for possible improvement. We sacrificed the optimality for simplicity in deriving the operator inequality. The definition of the phase error is not unique, and there may be a better choice, which will be sought in the next section. The trash rounds were introduced for technical reasons, but we are not sure whether they are really necessary. The protocols considered in Ref. [40, 41] use four or more states in signal or test modes, while ours use only two states. In fact, the binary protocol analyzed in Ref. [38] has the key rate comparable to ours when the excess noise ξ is around 10^{-3}.

In order to improve the presented finite-size key rate especially under the existence of excess noise, a promising route may be increasing the number of states from two. Our fidelity test can be straightforwardly generalized to the monitoring for such a larger constellation of signals, and we will be able to confine the adversary's attacks more tightly than in the present binary protocol. As for the proof techniques to determine the amount of privacy amplification, there are two possible directions. One is to generalize the present discrete-variable-QKD-inspired approach of estimating the number of phase errors in qubits to the case of qudits. The other direction is to seek a way to combine the existing analyses [40, 41, 83] of discrete-modulation continuous-variable QKD protocols, which have been reported to yield high key rates in the asymptotic regime, to our fidelity test. Although either of the approaches is nontrivial, we believe that the present results will open up a new direction toward exploiting the expected high potential of continuous-variable QKD with an improved security level.

4.5 Finite-Size Analysis for the Binary-Modulation Protocol Based on the Reverse Reconciliation

Although the security proof developed in the previous section, Sect. 4.4, realizes the finite-size security against general attacks in a composable fashion, the obtained key rate has very bad scaling against transmission distance. A possible reason for this bad performance is the way of security proof based on the entanglement distillation [61, 73]. There are alternative types of security proofs [49, 64, 66] that can approach necessary and sufficient conditions for security. In particular, for continuous-variable QKD protocols, it has been shown that the security proof based on the reverse reconciliation provides better performance than that based on the direct reconciliation [34].

In this section, we develop a refined security proof based on the reverse reconciliation (see Sect. 4.3 for its definition) for the binary-modulation protocol proposed in the previous section. With a refined proof and no additional experimental requirement, we obtain a significant improvement in the key gain against pure loss; in fact, it achieves near-optimal scaling against transmission distance in the asymptotic limit.

4.5.1 *Alternative Protocol*

Our refined security proof is for almost the same protocol as that in Sect. 4.4.2 but with a slight change. Prior to the protocol, Alice and Bob determine the number N of total rounds, the acceptance probability $f_{\mathrm{suc}}(q)$ ($q \in \mathbb{R}$) of the homodyne measurement satisfying $f_{\mathrm{suc}}(q) + f_{\mathrm{suc}}(-q) \leq 1$, the parameters for the test function (m, r), and the protocol parameters $(\mu, p_{\mathrm{sig}}, p_{\mathrm{test}}, p_{\mathrm{trash}}, \beta, s)$ with $p_{\mathrm{sig}} + p_{\mathrm{test}} + p_{\mathrm{trash}} = 1$, where all the parameters are positive. Alice and Bob then run the protocol described in the following.

Actual protocol

1. Alice generates a random bit $a \in \{0, 1\}$ and sends an optical pulse $\tilde{C}$ in a coherent state with an amplitude $(-1)^a \sqrt{\mu}$ to Bob. She repeats it N times.
2. For each of the received N pulses, Bob chooses a label from {signal, test, trash} with probabilities p_{sig}, p_{test}, and p_{trash}, respectively. According to the label, Alice and Bob do one of the following procedures.

   [signal] Bob performs a homodyne measurement on the received optical pulse C, and obtains an outcome $\check{q} \in \mathbb{R}$. Bob defines a bit b as $b = 0$ with a probability $f_{\text{suc}}(\check{q})$ and $b = 1$ with a probability $f_{\text{suc}}(-\check{q})$, and otherwise regards the round as "failure". The round in which Bob defines the bit b is regarded as "success". He announces success or failure of the detection. In the case of a success, Alice (resp. Bob) records a (b) as a sifted key bit.

   [test] Bob performs a heterodyne measurement on the received optical pulse C, and obtains an outcome $\check{\alpha}$. Alice announces her bit a. Bob calculates the value of $\Lambda_{m,r}(|\check{\alpha} - (-1)^a \beta|^2)$.

   [trash] Alice and Bob produce no outcomes.

3. We refer to the numbers of "success" and "failure" signal rounds, test rounds, and trash rounds as $\check{N}^{\text{suc}}$, $\check{N}^{\text{fail}}$, $\check{N}^{\text{test}}$, and $\check{N}^{\text{trash}}$, respectively. ($N = \check{N}^{\text{suc}} + \check{N}^{\text{fail}} + \check{N}^{\text{test}} + \check{N}^{\text{trash}}$ holds by definition. Note also that the sifted key length of this protocol is equal to $\check{N}^{\text{suc}}$.) Bob calculates the sum of $\Lambda_{m,r}(|\check{\alpha} - (-1)^a \beta|^2)$ obtained in the $\check{N}^{\text{test}}$ test rounds, which is denoted by $\check{F}$.
4. For the information reconciliation, they use $\check{N}_{\text{EC}}$-bits of encrypted communication consuming a pre-shared secret key to do the following. According to (the upper bound on) the bit error rate $\check{e}_{\text{ub}}$, Bob randomly chooses an error-correcting code and sends it with $\check{N}_{\text{EC}}$ bits of syndrome to Alice. Alice reconciles her sifted key accordingly.
5. Bob computes and announces the final key length $\check{N}^{\text{fin}}$ according to

$$\check{N}^{\text{fin}} = \max\left\{\check{N}^{\text{suc}} - \lceil \check{N}^{\text{suc}} h(\check{p}') \rceil - s, 0\right\}, \tag{4.223}$$

   where

$$\check{p}' := \min\left\{U'(\check{F}, \check{N}^{\text{trash}})/\check{N}^{\text{suc}}, 1/2\right\}. \tag{4.224}$$

   The function $U'(\check{F}, \check{N}^{\text{trash}})$ will be specified later. Alice and Bob apply the privacy amplification using the dual universal$_2$ hashing to obtain the final key.

For simplicity, we omitted the bit error sampling rounds in the above protocol. In order to estimate an upper bound $\check{e}_{\text{ub}}$ on the bit error rate with required correctness, Alice and Bob randomly insert N_{smp} sampling rounds among N rounds in which Bob performs the same measurement as that of the signal round. According to the observed discrepancies between Alice's and Bob's bits in the sampling rounds, Bob estimates the upper bound $\check{e}_{\mathrm{ub}}$ on the bit error rate with a failure probability no more than $\varepsilon_{\mathrm{cor}}/2$ and decides whether he aborts the protocol or not. The required amount of the error syndrome, $\check{N}_{\mathrm{EC}}$, that Bob sends to Alice in the information reconciliation step (Step 4) depends on the error correction method. Here we assume

$$\check{N}_{\mathrm{EC}} = \check{N}^{\mathrm{suc}}\left(fh(\check{e}_{\mathrm{ub}}) + (1-f)\right), \tag{4.225}$$

where $f \in [0, 1]$ denotes a reconciliation efficiency of an error-correcting code [40, 41, 43, 84–86] that succeeds with a probability no smaller than $1 - \varepsilon_{\mathrm{cor}}/2$, when given the upper bound $\check{e}_{\mathrm{ub}}$ on the bit error rate. (To satisfy the condition, the error-correcting code with which one can show an upper bound on the failure probability has to be used.) The net key gain $\check{G}$ per pulse is thus given by

$$\check{G} = (\check{N}^{\mathrm{fin}} - \check{N}_{\mathrm{EC}})/(N + N_{\mathrm{smp}}). \tag{4.226}$$

We do not use verification in the post-processing, unlike Sect. 4.4, due to subtleties to incorporate it in our security proof. The acceptance probability $f_{\mathrm{suc}}(q)$ should be chosen to post-select the rounds with larger values of q, for which the bit error probability is expected to be lower. Note that the encryption of the $\check{N}_{\mathrm{EC}}$-bit syndrome at the information reconciliation step can be replaced with the $\check{N}_{\mathrm{EC}}$-bit additional privacy amplification as long as the syndrome bits Bob sends to Alice in the protocol linearly depend on Bob's sifted key. See the previous section for a detailed explanation.

4.5.2 Security Proof Based on the Reverse Reconciliation

We determine a sufficient amount of the privacy amplification according to the complementarity summarized in Sects. 4.3.4 and 4.3.5 with the reverse reconciliation. To do so, we consider a virtual protocol in which Bob has a qubit for each success signal round such that the outcome of the Z-basis measurement on it coincides with his sifted key bit b. Alice can do arbitrary quantum operations in the virtual protocol as long as all the statistics and available information to the adversary Eve are the same as those in the actual protocol.

For Alice, we assume that a qubit A is introduced and the state in Eq. (4.117) is prepared in each round of the virtual protocol. Then, the optical quantum state emitted by Alice is the same as that in the actual protocol. For Bob, we adopt the same CP map $\mathcal{F}_{C\to B}$ defined in Eq. (4.118) to obtain a qubit B from the receive optical pulse C.

If the qubit B is successively measured on the Z basis, the obtained bit is equivalent to that in the signal round of the actual protocol.

At this point, one has the degree of freedom to perform quantum operations that do not change the value of the Z basis of the qubit B. Suppose that, after the qubit extraction map $\mathcal{F}_{C\to B}$ defined in Eq. (4.118), Alice and Bob performed a controlled-isometry $V_{BA\to BA'}$ that is defined as

$$V_{BA\to BA'} := |0\rangle\langle 0|_B \otimes V^{(0)}_{A\to A'} + |1\rangle\langle 1|_B \otimes V^{(1)}_{A\to A'}, \tag{4.227}$$

where $V^{(i)}_{A\to A'}$ denotes an isometry from the system A to another system A' that is no smaller than A.[9] (Note that this corresponds to the twisting operation on the shield system [63, 87–90] since what we will show in the following is equivalent to showing that the system B is private (i.e., secret) to Eve with the shield system A'.) We thus introduce a virtual protocol in the sense of Sect. 4.3.4 that explicitly incorporates the action of $V_{BA\to BA'}$ in the following.

Virtual protocol

1′. Alice prepares a qubit A and an optical pulse $\tilde{C}$ in a state $|\Psi\rangle_{A\tilde{C}}$ defined in (4.117). She repeats it N times.

2′. For each of the received N pulses, Bob announces a label in the same way as that in Step 2. Alice and Bob do one of the following procedures according to the label.

 [signal] Bob performs a quantum operation on the received pulse C specified by the CP map $\mathcal{F}_{C\to B}$ to determine success or failure of detection and to obtain qubit B upon success. He announces success or failure of detection. In the case of a success, Alice keeps her qubit A. Alice and Bob perform the controlled-isometry $V_{BA\to BA'}$ defined in Eq. (4.227).

 [test] Bob performs a heterodyne measurement on the received optical pulse C, and obtains an outcome $\check{\alpha}$. Alice measures her qubit A on Z basis and announces the outcome $a \in \{0, 1\}$. Bob calculates the value of $\Lambda_{m,r}(|\check{\alpha} - (-1)^a\beta|^2)$.

 [trash] Alice measures her qubit A on X basis to obtain $a' \in \{+, -\}$.

3′. $\check{N}^{\text{suc}}$, $\check{N}^{\text{fail}}$, $\check{N}^{\text{test}}$, $\check{N}^{\text{trash}}$, and $\check{F}$ are defined in the same way as those in Step 3. Let $\check{Q}_-$ be the number of rounds in the $\check{N}^{\text{trash}}$ trash rounds with $a' = -$.

4′. According to (the upper bound on) the bit error rate $\check{e}_{\text{ub}}$, Bob uses $\check{N}_{\text{EC}}$ bits of encrypted communication consuming a pre-shared secret key to send a dummy message to Alice.

5′. Bob computes and announces the final key length $\check{N}^{\text{fin}}$ according to Eq. (4.223). Bob acts a randomly chosen unitary on his qubits and measures the first $\check{N}^{\text{fin}}$ qubits on the Z bases.

To follow the security proof in Sects. 4.3.4 and 4.3.5, suppose that, at the end of Step 3′ in the virtual protocol, Bob measured his $\check{N}^{\text{suc}}$ qubits on the X basis $\{|+\rangle, |-\rangle\}$, and obtained a sequence $\check{x}$ of '+' and '−'. The '−' in $\check{x}$ is regarded as a phase error. If we can bound the number of possible phase-error patterns, then we can show the security

[9] Here, the subtleties for using the verification come in. In order to know whether verification succeeds or not, Alice has to confirm the syndrome bits for the verification. However, this confirmation procedure may not commute with the action of $V_{BA\to BA'}$. Therefore, unless we evaluate how much the verification affects the secrecy condition, we cannot use the verification in this security proof.

from the argument in Sect. 4.3.5. In order to make the argument more rigorous, we define an estimation protocol as follows.

Estimation protocol

$1'' - 3''$. Same as Steps $1'$, $2'$, and $3'$ of the virtual protocol.

$4''$. For every success signal round, Bob measures his qubit on the X basis and obtain an outcome $b' \in \{+, -\}$. Regarding $+$ as zero and $-$ as unity in the $\check{N}^{\mathrm{suc}}$ outcomes, define the $\check{N}^{\mathrm{suc}}$-bit sequence $\check{\boldsymbol{x}}$. Let $\check{N}_{\mathrm{ph}}^{\prime\mathrm{suc}} := \mathrm{wt}(\check{\boldsymbol{x}})$, where $\mathrm{wt}(\cdot)$ is defined in Eq. (4.4).

The task of proving the security of the actual protocol is then reduced to the construction of a function $U'(\check{F}, \check{N}^{\mathrm{trash}})$ that satisfies

$$\Pr\left[\check{N}_{\mathrm{ph}}^{\prime\mathrm{suc}} \le U'(\check{F}, \check{N}^{\mathrm{trash}})\right] \ge 1 - \epsilon \tag{4.228}$$

for any attack in the estimation protocol. In order to show that this is sufficient for the security proof, let $\mathcal{T}(\check{N}^{\mathrm{suc}}, \check{F}, \check{N}^{\mathrm{trash}})$ be the set of all the possible patterns for $\check{\boldsymbol{x}}$ with $\mathrm{wt}(\check{\boldsymbol{x}}) \le U'(\check{F}, \check{N}^{\mathrm{trash}})$. Then from Lemma 4.2.3, we have $\log|\mathcal{T}(\check{N}^{\mathrm{suc}}, \check{F}, \check{N}^{\mathrm{trash}})| \le N_{\mathrm{PA}}(\check{N}^{\mathrm{suc}}, \check{F}, \check{N}^{\mathrm{trash}})$ with

$$N_{\mathrm{PA}}(\check{N}^{\mathrm{suc}}, \check{F}, \check{N}^{\mathrm{trash}}) = \lceil \check{N}^{\mathrm{suc}} h(\check{p}') \rceil, \tag{4.229}$$

where $\check{p}'$ is defined in Eq. (4.224). With the same reasoning as the previous section, the condition (4.228) implies $\varepsilon_{\mathrm{sec}} = \sqrt{2(\epsilon + 2^{-s})}$-secrecy. Therefore, from now on, we focus on the estimation protocol.

4.5.3 Phase Error Operator

The number of phase errors depends on the choice of the controlled-isometry $V_{BA\to BA'}$ in the virtual and estimation protocol, and thus we aim to reduce the number of phase errors with a good choice of $V_{BA\to BA'}$. Here we heuristically choose $V_{BA\to BA'}$ so that the probability of the phase error event $b' = -$ in the estimation protocol is minimized for an ideal pure-loss channel (introduced in Sect. 3.1.3), which can be analytically investigated. We then use the same $V_{BA\to BA'}$ under the existence of excess noises, which is suboptimal but may still be a good choice when the excess noise is small.

When the state $|\Psi\rangle_{A\tilde{C}}$ in Eq. (4.117) is put into a pure-loss channel and the channel output is $|\pm\beta\rangle_C$, the resulting state $|\Phi\rangle_{ACE}$ on systems A, C, and an adversary's system E (i.e., environment of the pure-loss channel) is given by

$$|\Phi\rangle_{ACE} = \frac{1}{\sqrt{2}}\left(|0\rangle_A|\beta\rangle_C|\sqrt{\mu-\beta^2}\rangle_E + |1\rangle_A|-\beta\rangle_C|-\sqrt{\mu-\beta^2}\rangle_E\right). \tag{4.230}$$

Tracing out the system E, the reduced state Φ_{AC} is given by

$$\Phi_{AC} = \frac{e^{-(\mu-\beta^2)}}{2}\Big(\cosh(\mu-\beta^2)\,\hat{P}\big(|0\rangle_A\,|\beta\rangle_C + |1\rangle_A\,|-\beta\rangle_C\big) + \sinh(\mu-\beta^2)\,\hat{P}\big(|0\rangle_A\,|\beta\rangle_C - |1\rangle_A\,|-\beta\rangle_C\big)\Big) \tag{4.231}$$

$$= e^{-(\mu-\beta^2)}\Big(\cosh(\mu-\beta^2)\,\hat{P}\big(|+\rangle_A\,\Pi_{\mathrm{ev}}\,|\beta\rangle_C + |-\rangle_A\,\Pi_{\mathrm{od}}\,|\beta\rangle_C\big) + \sinh(\mu-\beta^2)\,\hat{P}\big(|+\rangle_A\,\Pi_{\mathrm{od}}\,|\beta\rangle_C + |-\rangle_A\,\Pi_{\mathrm{ev}}\,|\beta\rangle_C\big)\Big), \tag{4.232}$$

where $\hat{P}(\phi) := \phi\phi^\dagger$ (and thus $\hat{P}(|\phi\rangle) = |\phi\rangle\langle\phi|$). After Bob acts the qubit-extraction map $\mathcal{F}_{C\to B}$ defined in (4.118) to the state Φ_{AC} and obtains the subnormalized state $\tau_{AB} := \mathcal{F}_{C\to B}(\Phi_{AC})$ with $p_{\mathrm{suc}} := \mathrm{Tr}(\tau_{AB})$, he performs a controlled-isometry $V_{BA\to BA'}$ on τ_{AB}. The phase error probability can thus be given by

$$\mathrm{Tr}[|-\rangle\langle-|_B\, V_{BA\to BA'}\tau_{AB}V^\dagger_{BA\to BA'}] \tag{4.233}$$

$$= \frac{1}{2}\mathrm{Tr}\Big[\langle 0|_B\,\tau_{AB}\,|0\rangle_B + \langle 1|_B\,\tau_{AB}\,|1\rangle_B - V^{(1)\dagger}_{A\to A'}V^{(0)}_{A\to A'}\,\langle 0|_B\,\tau_{AB}\,|1\rangle_B - \langle 1|_B\,\tau_{AB}\,|0\rangle_B\,V^{(1)}_{A\to A'}V^{(0)\dagger}_{A\to A'}\Big] \tag{4.234}$$

$$= \frac{1}{2}p_{\mathrm{suc}} - \mathrm{Re}\Big[\mathrm{Tr}\Big(V^{(1)\dagger}_{A\to A'}V^{(0)}_{A\to A'}\,\langle 0|_B\,\tau_{AB}\,|1\rangle_B\Big)\Big] \tag{4.235}$$

$$\geq \frac{1}{2}p_{\mathrm{suc}} - \|\,\langle 0|_B\,\tau_{AB}\,|1\rangle_B\,\|_1, \tag{4.236}$$

where $V^{(i)}_{A\to A'}$ $(i = 0, 1)$ are defined in Eq. (4.227), and the last inequality follows from the matrix Hölder inequality. If we write the polar decomposition of $\langle 0|_B\,\tau_{AB}\,|1\rangle_B$ by $W_A\,\big|\langle 0|_B\,\tau_{AB}\,|1\rangle_B\big|$, the equality in (4.236) can be achieved by setting $V^{(1)\dagger}_{A\to A'}V^{(0)}_{A\to A'} = W^\dagger_A$. One can also check that $\langle 0|_B\,\tau_{AB}\,|1\rangle_B$ is a real matrix on the Z basis, and therefore, W_A can be taken to be an orthogonal matrix on the Z basis (and thus on the X basis).

We will derive an explicit form of W_A as a function of μ and β. We first observe that

$$\sigma^X_A\,\langle 0|_B\,\tau_{AB}\,|1\rangle_B = \frac{1}{2\sqrt{\pi}}\int_{-\infty}^{\infty} f_{\mathrm{suc}}(q)dq\,\Big[e^{-2(\mu-\beta^2)}\big(e^{-(q+\sqrt{2}\beta)^2}\,|0\rangle\langle 0|_A + e^{-(q-\sqrt{2}\beta)^2}\,|1\rangle\langle 1|_A\big) + e^{-q^2-2\beta^2}\big(|1\rangle\langle 0|_A + |0\rangle\langle 1|_A\big)\Big] \tag{4.237}$$

$$= \frac{1}{2}\big[a(\mu,\beta)\,|0\rangle\langle 0|_A + b(\mu,\beta)\,|1\rangle\langle 1|_A + c(\beta)\,\big(|1\rangle\langle 0|_A + |0\rangle\langle 1|_A\big)\big] \tag{4.238}$$

with real positive functions $a(\mu,\beta)$, $b(\mu,\beta)$, and $c(\beta)$ defined as

$$a(\mu, \beta) = e^{-2(\mu-\beta^2)} G_{\frac{1}{2}} * f_{\mathrm{suc}}(-\sqrt{2}\beta), \tag{4.239}$$

$$b(\mu, \beta) = e^{-2(\mu-\beta^2)} G_{\frac{1}{2}} * f_{\mathrm{suc}}(\sqrt{2}\beta), \tag{4.240}$$

$$c(\beta) = e^{-2\beta^2} G_{\frac{1}{2}} * f_{\mathrm{suc}}(0), \tag{4.241}$$

where $G_{\frac{1}{2}}(q) := \exp(-q^2)/\sqrt{\pi}$ denotes the normal distribution with variance 1/2, and $f * g$ denotes the convolution of the functions f and g. The matrix $\sigma_A^X \langle 0|_B \tau_{AB} |1\rangle_B$ is thus hermitian. If $a(\mu, \beta)b(\mu, \beta) - c(\beta)^2 \geq 0$ or equivalently $\mu \leq \mu_{\mathrm{th}}(\beta)$ with

$$\mu_{\mathrm{th}}(\beta) := 2\beta^2 - \frac{1}{4} \ln \left(\frac{\left[G_{\frac{1}{2}} * f_{\mathrm{suc}}(0)\right]^2}{\left[G_{\frac{1}{2}} * f_{\mathrm{suc}}(\sqrt{2}\beta)\right]\left[G_{\frac{1}{2}} * f_{\mathrm{suc}}(-\sqrt{2}\beta)\right]} \right), \tag{4.242}$$

then the matrix $\sigma_A^X \langle 0|_B \tau_{AB} |1\rangle_B$ is positive, which leads us to setting $W_A = \sigma_A^X$. When $W_A = \sigma_A^X$, the controlled-isometry $V_{BA\to BA'}$ is chosen to be the CNOT, and thus the security proof is completely reduced to the previous analysis in Sect. 4.4 with the operator inequality (4.255) given by Corollary 4.4.4. On the other hand, when $\mu > \mu_{\mathrm{th}}(\beta)$ (i.e., $a(\mu, \beta)b(\mu, \beta) - c(\beta)^2 < 0$), we rewrite Eq. (4.238) as

$$\sigma_A^X \langle 0|_B \tau_{AB} |1\rangle_B = \frac{D(\mu, \beta)}{4} \left[\frac{a(\mu, \beta) + b(\mu, \beta)}{D(\mu, \beta)} I_A + \cos\theta(\mu, \beta)\, \sigma_A^X + \sin\theta(\mu, \beta)\, \sigma_A^Z \right], \tag{4.243}$$

where $D(\mu, \beta) := \sqrt{[a(\mu, \beta) - b(\mu, \beta)]^2 + 4c(\beta)^2} (> 0)$, $[a(\mu, \beta) + b(\mu, \beta)]/D(\mu, \beta) < 1$, and $\theta(\mu, \beta)$ satisfies

$$|\theta(\mu, \beta)| < \pi/2 \quad \text{and} \quad \tan\theta(\mu, \beta) = \frac{a(\mu, \beta) - b(\mu, \beta)}{2c(\beta)}. \tag{4.244}$$

It shows that we may choose W_A as

$$W_A = \sigma_A^X \left(\cos\theta(\mu, \beta)\sigma_A^X + \sin\theta(\mu, \beta)\sigma_A^Z\right) = \cos\theta(\mu, \beta) I_A - i \sin\theta(\mu, \beta)\sigma_A^Y \tag{4.245}$$

because

$$\begin{aligned} &W_A^\dagger \langle 0|_B \tau_{AB} |1\rangle_B \\ &= \frac{a(\mu, \beta) + b(\mu, \beta)}{4} \left[\frac{D(\mu, \beta)}{a(\mu, \beta) + b(\mu, \beta)} I_A + \cos\theta(\mu, \beta)\, \sigma_A^X + \sin\theta(\mu, \beta)\, \sigma_A^Z \right] \end{aligned} \tag{4.246}$$

is positive. From Eq. (4.244), the condition $\theta(\mu, \beta) = 0$ and thus $W_A = I_A$ holds only when $a(\mu, \beta) = b(\mu, \beta)$, i.e.,

$$G_{\frac{1}{2}} * f_{\mathrm{suc}}(\sqrt{2}\beta) = G_{\frac{1}{2}} * f_{\mathrm{suc}}(-\sqrt{2}\beta), \tag{4.247}$$

which is false for reasonable choices of the acceptance probability $f_{\rm suc}(q)$ that post-select larger values of q. Thus, in the following, we consider the case $\mu > \mu_{\rm th}(\beta)$ and $0 < |\theta(\mu, \beta)| < \pi/2$.

As we explained previously, we set $V^{(1)\dagger}_{A\to A'} V^{(0)}_{A\to A'} = W^\dagger_A$ also for arbitrary channels, i.e., arbitrary coherent attacks by Eve. This choice of $V_{BA\to BA'}$ is not optimal for general channels but is expected to be close to the optimal for channels that are close to the pure loss. Now that the controlled-isometry $V_{BA\to BA'}$ is fixed, we can clarify using Eq. (4.233) what combination of measurement events on Alice's qubit A and the optical pulse C leads to the phase error $b' = -$ in the estimation protocol. The operator $M'^{\rm suc}_{\rm ph}$ on the system A and C that corresponds to the phase error event is given by

$$M'^{\rm suc}_{\rm ph} := \mathcal{F}^\dagger_{C\to B}\big(V^\dagger_{BA\to BA'}(|-\rangle\langle -|_B \otimes I_{A'})V_{BA\to BA'}\big), \tag{4.248}$$

where $\mathcal{F}^\dagger_{C\to B}$ denotes the adjoint map of $\mathcal{F}_{C\to B}$ given explicitly by

$$\mathcal{F}^\dagger_{C\to B}(M_B) = \int_{-\infty}^{\infty} dq\, K^{(q)\dagger} M_B K^{(q)}. \tag{4.249}$$

Using the expression (4.119) of $K^{(q)}$, we have

$$M'^{\rm suc}_{\rm ph} = \frac{1}{2}\mathcal{F}^\dagger_{C\to B}\Big(\hat{P}\big(V^{(0)\dagger}_{A\to A'} \otimes |0\rangle_B - V^{(1)\dagger}_{A\to A'} \otimes |1\rangle_B\big)\Big) \tag{4.250}$$

$$= \frac{1}{2}\mathcal{F}^\dagger_{C\to B}\Big(I_{AB} - W_A \otimes |0\rangle\langle 1|_B - W^\dagger_A \otimes |1\rangle\langle 0|_B\Big) \tag{4.251}$$

$$= \frac{1}{4}\mathcal{F}^\dagger_{C\to B}\Big(\hat{P}\big(I_A \otimes (|+\rangle_B + |-\rangle_B) - W^\dagger_A \otimes (|+\rangle_B - |-\rangle_B)\big)\Big) \tag{4.252}$$

$$= \int_{-\infty}^{\infty} 2 f_{\rm suc}(q) dq\, \hat{P}\Big(\frac{I_A - W^\dagger_A}{2} \otimes \Pi_{\rm ev}\, |q\rangle_C + \frac{I_A + W^\dagger_A}{2} \otimes \Pi_{\rm od}\, |q\rangle_C\Big). \tag{4.253}$$

Once the phase error operator can be defined on systems A and C, we can follow essentially the same analysis as that in Sect. 4.4, replacing $M^{\rm suc}_{\rm ph}$ with $M'^{\rm suc}_{\rm ph}$. This also leads to replacing $M[\kappa, \gamma]$ with $M'[\kappa, \gamma]$ defined as

$$M'[\kappa, \gamma] := M'^{\rm suc}_{\rm ph} + \kappa \Pi^{\rm fid} - \gamma \Pi^{\rm trash}_{-}, \tag{4.254}$$

where Π^{fid} and Π_{-}^{trash} are defined in Eqs. (4.131) and (4.132). Then, the only difference to proceed the security proof is the operator inequality (4.140); we alternatively need to show

$$M'[\kappa, \gamma] \le B'(\kappa, \gamma) I_{AC}, \tag{4.255}$$

with a computable function $B'(\kappa, \gamma)$. Equation (4.255) will be shown in the next section.

Once we obtain $B'(\kappa, \gamma)$, following the same line of argument as that in Sect. 4.4 leads to the definition of

$$U'(\check{F}, \check{N}^{\mathrm{trash}}) = p_{\mathrm{sig}}\big(N B'(\kappa, \gamma) + \delta_1(\epsilon/2)\big) - \frac{p_{\mathrm{sig}}}{p_{\mathrm{test}}}\kappa \check{F} + \frac{p_{\mathrm{sig}}}{p_{\mathrm{trash}}}\gamma\Big(\check{N}^{\mathrm{trash}} q_{-} + \delta_2(\epsilon/2; \check{N}^{\mathrm{trash}})\Big), \tag{4.256}$$

which achieves $\sqrt{2(\epsilon + 2^{-s})}$-secrecy. Thus, we complete the finite-size security proof based on the reverse reconciliation for this alternative protocol.

4.5.4 Proof of the Operator Inequality

In this section, we prove the inequality (4.255) used in the security proof. We first prove the following lemma.

Lemma 4.5.1 *Let $\Pi_{\pm}$ be orthogonal projections which have rank no smaller than three or infinite. Let M be a self-adjoint operator satisfying $M = (\Pi_{+} + \Pi_{-}) M (\Pi_{+} + \Pi_{-}) \le \alpha(\Pi_{+} + \Pi_{-})$, where α is a real constant. Let $|\psi\rangle$ be a vector satisfying $(\Pi_{+} + \Pi_{-})|\psi\rangle = |\psi\rangle$ and $\Pi_{\pm}|\psi\rangle \ne 0$. Assume $\Pi_{+}|\psi\rangle$ is not proportional to the eigenvectors of $\Pi_{\pm} M \Pi_{+}$ (if they have). Define the following quantities with respect to $|\psi\rangle$:*

$$C_{\pm} := \langle\psi| \Pi_{\pm} |\psi\rangle \ (> 0), \tag{4.257}$$

$$\lambda_{\pm\pm} := C_{\pm}^{-1} \langle\psi| M_{\pm\pm} |\psi\rangle, \tag{4.258}$$

$$\lambda_{+-} := (C_{+}C_{-})^{-\frac{1}{2}} \langle\psi| M_{+-} |\psi\rangle, \qquad \lambda_{-+} := \lambda_{+-}^{*}, \tag{4.259}$$

$$\sigma_{\pm+} := \left(C_{+}^{-1} \| M_{\pm+} |\psi\rangle \|^2 - |\lambda_{\pm+}|^2\right)^{\frac{1}{2}} \ (> 0), \tag{4.260}$$

$$\sigma_{\pm-} := \sigma_{\pm+}^{-1} \left((C_{+}C_{-})^{-\frac{1}{2}} \langle\psi| M_{+\pm} M_{\pm-} |\psi\rangle - \lambda_{+-}\lambda_{\pm\pm}\right), \tag{4.261}$$

$$\Delta_{\pm-} := \left(C_{-}^{-1} \| M_{\pm-} |\psi\rangle \|^2 - |\lambda_{\pm-}|^2 - |\sigma_{\pm-}|^2\right)^{\frac{1}{2}}, \tag{4.262}$$

where M_{++}, M_{--}, M_{+-}, and M_{-+} are given respectively by

$$M_{\pm\pm} := \Pi_{\pm} M \Pi_{\pm}, \quad M_{+-} := \Pi_{+} M \Pi_{-}, \quad M_{-+} := M_{+-}^{\dagger}. \tag{4.263}$$

Then, for any real numbers $\gamma_{\pm}$, we have

$$\lambda_{\sup}(M + |\psi\rangle\langle\psi| - \gamma_+\Pi_+ - \gamma_-\Pi_-) \leq \lambda_{\sup}(M_{6d}), \tag{4.264}$$

where M_{6d} is given by

$$M_{6d} := \begin{pmatrix} \alpha-\gamma_+ & 0 & 0 & \Delta_{+-} & 0 & 0 \\ 0 & \alpha-\gamma_+ & \sigma_{++} & \sigma_{+-} & 0 & 0 \\ 0 & \sigma_{++} & C_+ + \lambda_{++} - \gamma_+ & \sqrt{C_+C_-} + \lambda_{+-} & \sigma_{-+} & 0 \\ \Delta_{+-} & \sigma^*_{+-} & \sqrt{C_+C_-} + \lambda_{-+} & C_- + \lambda_{--} - \gamma_- & \sigma^*_{--} & \Delta_{--} \\ 0 & 0 & \sigma_{-+} & \sigma_{--} & \alpha-\gamma_- & 0 \\ 0 & 0 & 0 & \Delta_{--} & 0 & \alpha-\gamma_- \end{pmatrix}. \tag{4.265}$$

Proof We choose orthonormal vectors $\{|e^{(1)}_\pm\rangle, |e^{(2)}_\pm\rangle, |e^{(3)}_\pm\rangle\}$ in the domains of $\Pi_\pm$, respectively, to satisfy

$$\sqrt{C_\pm}\,|e^{(1)}_\pm\rangle = \Pi_\pm|\psi\rangle, \tag{4.266}$$

$$M|e^{(1)}_+\rangle = (M_{++} + M_{-+})|e^{(1)}_+\rangle = \lambda_{++}|e^{(1)}_+\rangle + \sigma_{++}|e^{(2)}_+\rangle + \lambda_{-+}|e^{(1)}_-\rangle + \sigma_{-+}|e^{(2)}_-\rangle, \tag{4.267}$$

$$M|e^{(1)}_-\rangle = (M_{+-} + M_{--})|e^{(1)}_-\rangle = \lambda_{+-}|e^{(1)}_+\rangle + \sigma_{+-}|e^{(2)}_+\rangle + \Delta_{+-}|e^{(3)}_+\rangle \tag{4.268}$$

$$+ \lambda_{--}|e^{(1)}_-\rangle + \sigma_{--}|e^{(2)}_-\rangle + \Delta_{--}|e^{(3)}_-\rangle, \tag{4.269}$$

which is well-defined due to Eqs. (4.257)–(4.263) and $M = (\Pi_+ + \Pi_-)M(\Pi_+ + \Pi_-)$. Actually, Eqs. (4.260)–(4.262) are derived by taking inner product of appropriate pairs among $M_{\pm\pm}|\psi\rangle$ and $M_{\pm\mp}|\psi\rangle$. Overall phases of $|e^{(2)}_\pm\rangle$ and $|e^{(3)}_\pm\rangle$ are taken so that $\sigma_{\pm+}$ and $\Delta_{\pm-}$ are non-negative. Since $\Pi_+|\psi\rangle$ is not proportional to the eigenvectors of $\Pi_\pm M\Pi_+$ by assumption, we have $\sigma_{\pm+} > 0$. From $(\Pi_+ + \Pi_-)|\psi\rangle = |\psi\rangle$, we have

$$|\psi\rangle = \sqrt{C_+}\,|e^{(1)}_+\rangle + \sqrt{C_-}\,|e^{(1)}_-\rangle. \tag{4.270}$$

Let us now define following projection operators:

$$\Pi^{(j)}_\pm := |e^{(j)}_\pm\rangle\langle e^{(j)}_\pm| \quad (j = 1, 2, 3), \tag{4.271}$$

$$\Pi^{(\geq 2)}_\pm := \Pi_\pm - \Pi^{(1)}_\pm, \tag{4.272}$$

$$\Pi^{(\geq 4)}_\pm := \Pi^{(\geq 2)}_\pm - \Pi^{(2)}_\pm - \Pi^{(3)}_\pm. \tag{4.273}$$

Since Eqs. (4.269) and (4.269) imply $(\Pi^{(\geq 4)}_+ + \Pi^{(\geq 4)}_-)M(\Pi^{(1)}_+ + \Pi^{(1)}_-) = 0$, we have

$$M = (\Pi_+ + \Pi_-)M(\Pi_+ + \Pi_-) \tag{4.274}$$

$$= (\Pi^{(1)}_+ + \Pi^{(1)}_-)M(\Pi^{(1)}_+ + \Pi^{(1)}_-) + (\Pi^{(2)}_+ + \Pi^{(3)}_+ + \Pi^{(2)}_- + \Pi^{(3)}_-)M(\Pi^{(1)}_+ + \Pi^{(1)}_-)$$

$$+ (\Pi_+^{(1)} + \Pi_-^{(1)}) M (\Pi_+^{(2)} + \Pi_+^{(3)} + \Pi_-^{(2)} + \Pi_-^{(3)}) + (\Pi_+^{(\geq 2)} + \Pi_-^{(\geq 2)}) M (\Pi_+^{(\geq 2)} + \Pi_-^{(\geq 2)}) \tag{4.275}$$

$$\begin{aligned} &\leq \lambda_{++}\Pi_+^{(1)} + \lambda_{--}\Pi_-^{(1)} + \lambda_{+-} |e_+^{(1)}\rangle\langle e_-^{(1)}| + \lambda_{-+} |e_-^{(1)}\rangle\langle e_+^{(1)}| \\ &+ \Big(\sigma_{++} |e_+^{(2)}\rangle\langle e_+^{(1)}| + \sigma_{-+} |e_-^{(2)}\rangle\langle e_+^{(1)}| + \sigma_{+-} |e_+^{(2)}\rangle\langle e_-^{(1)}| \\ &\quad + \Delta_{+-} |e_+^{(3)}\rangle\langle e_-^{(1)}| + \sigma_{--} |e_-^{(2)}\rangle\langle e_-^{(1)}| + \Delta_{--} |e_-^{(3)}\rangle\langle e_-^{(1)}|\Big) + (\text{h.c.}) \\ &+ \alpha(\Pi_+^{(\geq 2)} + \Pi_-^{(\geq 2)}), \end{aligned} \tag{4.276}$$

where h.c. denotes the hermitian conjugates of the terms in the preceding parenthesis. The last inequality comes from $M \leq \alpha(\Pi_+ + \Pi_-)$. Using Eq. (4.276), we have

$$M + |\psi\rangle\langle\psi| - \gamma_+\Pi_+ - \gamma_-\Pi_- \leq M_{6\text{d}} \oplus (\alpha - \gamma_+)\Pi_+^{(\geq 4)} \oplus (\alpha - \gamma_-)\Pi_-^{(\geq 4)}, \tag{4.277}$$

where $M_{6\text{d}}$ is given in Eq. (4.265) with the basis $\{|e_+^{(3)}\rangle, |e_+^{(2)}\rangle, |e_+^{(1)}\rangle, |e_-^{(1)}\rangle, |e_-^{(2)}\rangle, |e_-^{(3)}\rangle\}$. Since $\alpha - \gamma_\pm = \langle e_\pm^{(3)}| M_{6\text{d}} |e_\pm^{(3)}\rangle \leq \lambda_{\sup}(M_{6\text{d}})$, the supremum of the spectrum of the right-hand side of Eq. (4.277) is equal to the maximum eigenvalue of the six-dimensional matrix $M_{6\text{d}}$. We then obtain Eq. (4.264). □

As a corollary of this lemma, we obtain the following.

Corollary 4.5.2 *Let $|\beta\rangle$ be a coherent state. Let $\mu_{\text{th}}(\beta)$ and $\theta(\mu, \beta)$ be as defined in Eqs. (4.242) and (4.244), respectively. Suppose $\mu > \mu_{\text{th}}(\beta)$ and $\theta(\mu, \beta) \neq 0$. Let $\Pi_{\text{ev(od)}}$ and $M'[\kappa, \gamma]$ be as defined in Sect. 4.5.3 with W_A given in Eq. (4.245). Let $M_{\text{ee}}^{\text{suc}}$, $M_{\text{oo}}^{\text{suc}}$, $M_{\text{eo}}^{\text{suc}}$, and $M_{\text{oe}}^{\text{suc}}$ be defined as follows:*

$$M_{\text{ee}}^{\text{suc}} := \int_{-\infty}^{\infty} 2 f_{\text{suc}}(q) dq\, \Pi_{\text{ev}} |q\rangle\langle q| \Pi_{\text{ev}}, \tag{4.278}$$

$$M_{\text{oo}}^{\text{suc}} := \int_{-\infty}^{\infty} 2 f_{\text{suc}}(q) dq\, \Pi_{\text{od}} |q\rangle\langle q| \Pi_{\text{od}}, \tag{4.279}$$

$$M_{\text{eo}}^{\text{suc}} := \int_{-\infty}^{\infty} 2 f_{\text{suc}}(q) dq\, \Pi_{\text{ev}} |q\rangle\langle q| \Pi_{\text{od}}, \tag{4.280}$$

$$M_{\text{oe}}^{\text{suc}} := M_{\text{eo}}^{\text{suc}\dagger}. \tag{4.281}$$

Define the following (real) parameters:

$$w_+ := \frac{1 + \cos\theta(\mu, \beta)}{2}, \quad w_- := \frac{1 - \cos\theta(\mu, \beta)}{2}, \quad w_z := \frac{\sin\theta(\mu, \beta)}{2}, \tag{4.282}$$

$$C_{\text{o}} := \langle\beta| \Pi_{\text{od}} |\beta\rangle = e^{-|\beta|^2} \sinh|\beta|^2, \quad C_{\text{e}} := \langle\beta| \Pi_{\text{ev}} |\beta\rangle = e^{-|\beta|^2} \cosh|\beta|^2, \tag{4.283}$$

$$\lambda_{oo} := \frac{w_+}{C_o} \langle\beta| M_{oo}^{suc} |\beta\rangle, \quad \lambda_{ee} := \frac{w_-}{C_e} \langle\beta| M_{ee}^{suc} |\beta\rangle, \tag{4.284}$$

$$\lambda_{oe} := \frac{w_z}{\sqrt{C_o C_e}} \langle\beta| M_{oe}^{suc} |\beta\rangle, \quad \lambda_{eo} := \frac{w_z}{\sqrt{C_o C_e}} \langle\beta| M_{eo}^{suc} |\beta\rangle = \lambda_{oe}^*, \tag{4.285}$$

$$\sigma_{oo} := \left(\frac{w_+^2}{C_o} \| M_{oo}^{suc} |\beta\rangle \|^2 - \lambda_{oo}^2\right)^{\frac{1}{2}}, \quad \sigma_{eo} := \left(\frac{w_z^2}{C_o} \| M_{eo}^{suc} |\beta\rangle \|^2 - \lambda_{eo}^2\right)^{\frac{1}{2}}, \tag{4.286}$$

$$\sigma_{oe} := \sigma_{oo}^{-1} \left(\frac{w_+ w_z}{\sqrt{C_o C_e}} \langle\beta| M_{oo}^{suc} M_{oe}^{suc} |\beta\rangle - \lambda_{oo}\lambda_{oe}\right), \tag{4.287}$$

$$\sigma_{ee} := \sigma_{eo}^{-1} \left(\frac{w_- w_z}{\sqrt{C_o C_e}} \langle\beta| M_{oe}^{suc} M_{ee}^{suc} |\beta\rangle - \lambda_{oe}\lambda_{ee}\right), \tag{4.288}$$

$$\Delta_{oe} := \left(\frac{w_z^2}{C_e} \| M_{oe}^{suc} |\beta\rangle \|^2 - \lambda_{oe}^2 - \sigma_{oe}^2\right)^{\frac{1}{2}}, \quad \Delta_{ee} := \left(\frac{w_-^2}{C_e} \| M_{ee}^{suc} |\beta\rangle \|^2 - \lambda_{ee}^2 - \sigma_{ee}^2\right)^{\frac{1}{2}}. \tag{4.289}$$

Define the following two matrices $M_{6d}^{(0)}$ and $M_{6d}^{(1)}$.

$$M_{6d}^{(0)} := \begin{pmatrix} 1 & 0 & 0 & \Delta_{oe} & 0 & 0 \\ 0 & 1 & \sigma_{oo} & \sigma_{oe} & 0 & 0 \\ 0 & \sigma_{oo} & \kappa C_o + \lambda_{oo} & \kappa\sqrt{C_o C_e} + \lambda_{oe} & \sigma_{eo} & 0 \\ \Delta_{oe} & \sigma_{oe} & \kappa\sqrt{C_o C_e} + \lambda_{eo} & \kappa C_e + \lambda_{ee} - \gamma & \sigma_{ee} & \Delta_{ee} \\ 0 & 0 & \sigma_{eo} & \sigma_{ee} & 1-\gamma & 0 \\ 0 & 0 & 0 & \Delta_{ee} & 0 & 1-\gamma \end{pmatrix}, \tag{4.290}$$

$$M_{6d}^{(1)} := \begin{pmatrix} 1-\gamma & 0 & 0 & \Delta_{oe} & 0 & 0 \\ 0 & 1-\gamma & \sigma_{oo} & -\sigma_{oe} & 0 & 0 \\ 0 & \sigma_{oo} & \kappa C_o + \lambda_{oo} - \gamma & \kappa\sqrt{C_o C_e} - \lambda_{oe} & \sigma_{eo} & 0 \\ \Delta_{oe} & -\sigma_{oe} & \kappa\sqrt{C_o C_e} - \lambda_{eo} & \kappa C_e + \lambda_{ee} & -\sigma_{ee} & \Delta_{ee} \\ 0 & 0 & \sigma_{eo} & -\sigma_{ee} & 1 & 0 \\ 0 & 0 & 0 & \Delta_{ee} & 0 & 1 \end{pmatrix}. \tag{4.291}$$

Define a convex function

$$B'(\kappa, \gamma) := \max\{\lambda_{sup}(M_{6d}^{(0)}), \lambda_{sup}(M_{6d}^{(1)})\}. \tag{4.292}$$

Then, for $\kappa, \gamma \geq 0$, we have

$$M'[\kappa, \gamma] \leq B'(\kappa, \gamma) I_{AC}. \tag{4.293}$$

Proof From the form of W_A in Eq. (4.245), we have

$$\left\| \frac{I+W_A}{2}\left|+\right\rangle \right\|^2 = \left\| \frac{I+W_A}{2}\left|-\right\rangle \right\|^2 = \frac{1+\cos\theta(\mu,\beta)}{2} = w_+, \tag{4.294}$$

$$\left\| \frac{I-W_A}{2}\left|-\right\rangle \right\|^2 = \left\| \frac{I-W_A}{2}\left|+\right\rangle \right\|^2 = \frac{1-\cos\theta(\mu,\beta)}{2} = w_-, \tag{4.295}$$

$$\begin{aligned} &-\left\langle -\right| \frac{I+W_A^\dagger}{2}\frac{I-W_A}{2}\left|+\right\rangle = -\left\langle +\right| \frac{I-W_A^\dagger}{2}\frac{I+W_A}{2}\left|-\right\rangle \\ &= \left\langle +\right| \frac{I+W_A^\dagger}{2}\frac{I-W_A}{2}\left|-\right\rangle = \left\langle -\right| \frac{I-W_A^\dagger}{2}\frac{I+W_A}{2}\left|+\right\rangle = \frac{\sin\theta(\mu,\beta)}{2} = w_z, \end{aligned} \tag{4.296}$$

where $w_\pm$ are positive due to $0 < |\theta(\mu,\beta)| < \pi/2$. Since $\sigma_A^X W_A = W_A^\dagger \sigma_A^X$ holds, the operator acted by $\mathcal{F}_{C\to B}^\dagger(\cdot)$ in Eq. (4.251) commutes with $\sigma_A^X \otimes \sigma_B^X$. Due to the form of $\mathcal{F}_{C\to B}^\dagger(\cdot)$, this means that $M_{\mathrm{ph}}^{\prime\mathrm{suc}}$ commutes with $\sigma_A^X \otimes \exp(\pi i \hat{n}_C)$, and so does $M'[\kappa,\gamma]$. Therefore, we have

$$M'[\kappa,\gamma] = \Pi_{AC}^{(+,\mathrm{od}),(-,\mathrm{ev})} M'[\kappa,\gamma] \Pi_{AC}^{(+,\mathrm{od}),(-,\mathrm{ev})} + \Pi_{AC}^{(-,\mathrm{od}),(+,\mathrm{ev})} M'[\kappa,\gamma] \Pi_{AC}^{(-,\mathrm{od}),(+,\mathrm{ev})}, \tag{4.297}$$

where two orthogonal projections $\Pi_{AC}^{(+,\mathrm{od}),(-,\mathrm{ev})}$ and $\Pi_{AC}^{(-,\mathrm{od}),(+,\mathrm{ev})}$ are defined as

$$\Pi_{AC}^{(+,\mathrm{od}),(-,\mathrm{ev})} := \left|+\right\rangle\left\langle +\right|_A \otimes \Pi_{\mathrm{od}} + \left|-\right\rangle\left\langle -\right|_A \otimes \Pi_{\mathrm{ev}} \tag{4.298}$$

$$\Pi_{AC}^{(-,\mathrm{od}),(+,\mathrm{ev})} := \left|-\right\rangle\left\langle -\right|_A \otimes \Pi_{\mathrm{od}} + \left|+\right\rangle\left\langle +\right|_A \otimes \Pi_{\mathrm{ev}}. \tag{4.299}$$

Then we apply Lemma 4.5.1 to the operators $\Pi_{AC}^{(+,\mathrm{od}),(-,\mathrm{ev})} M'[\kappa,\gamma] \Pi_{AC}^{(+,\mathrm{od}),(-,\mathrm{ev})}$ and $\Pi_{AC}^{(-,\mathrm{od}),(+,\mathrm{ev})} M'[\kappa,\gamma] \Pi_{AC}^{(-,\mathrm{od}),(+,\mathrm{ev})}$, respectively. For $\Pi_{AC}^{(+,\mathrm{od}),(-,\mathrm{ev})} M'[\kappa,\gamma]$ $\Pi_{AC}^{(+,\mathrm{od}),(-,\mathrm{ev})}$, we set

$$\Pi_\pm = \left|\pm\right\rangle\left\langle \pm\right|_A \otimes \Pi_{\mathrm{od(ev)}}, \tag{4.300}$$

$$M = \Pi_{AC}^{(+,\mathrm{od}),(-,\mathrm{ev})} M_{\mathrm{ph}}^{\mathrm{suc}} \Pi_{AC}^{(+,\mathrm{od}),(-,\mathrm{ev})} \tag{4.301}$$

$$= w_+ \left|+\right\rangle\left\langle +\right| \otimes M_{\mathrm{oo}}^{\mathrm{suc}} + w_- \left|-\right\rangle\left\langle -\right| \otimes M_{\mathrm{ee}}^{\mathrm{suc}} + w_z \left(\left|+\right\rangle\left\langle -\right| \otimes M_{\mathrm{oe}}^{\mathrm{suc}} + \left|-\right\rangle\left\langle +\right| \otimes M_{\mathrm{eo}}^{\mathrm{suc}}\right), \tag{4.302}$$

$$\left|\psi\right\rangle = \sqrt{\kappa}\left|\phi_{\mathrm{err}}\right\rangle_{AC}, \tag{4.303}$$

$$\alpha = 1, \quad \gamma_+ = 0, \quad \gamma_- = \gamma, \tag{4.304}$$

where $\left|\phi_{\mathrm{err}}\right\rangle_{AC}$ is defined in Eq. (4.190). Since so-defined M only has continuous spectrum, we can apply Lemma 4.5.1 and obtain

$$\lambda_{\sup}\left(\Pi_{AC}^{(+,\mathrm{od}),(-,\mathrm{ev})} M'[\kappa,\gamma] \Pi_{AC}^{(+,\mathrm{od}),(-,\mathrm{ev})}\right) \le \lambda_{\sup}(M_{\mathrm{6d}}^{(0)}). \tag{4.305}$$

In the same way, we apply Lemma 4.5.1 to $\Pi_{AC}^{(-,\mathrm{od}),(+,\mathrm{ev})} M'[\kappa,\gamma] \Pi_{AC}^{(-,\mathrm{od}),(+,\mathrm{ev})}$ by setting

$$\Pi_{\pm} = |\mp\rangle\langle\mp|_A \otimes \Pi_{\mathrm{od(ev)}}, \tag{4.306}$$

$$M = \Pi_{AC}^{(-,\mathrm{od}),(+,\mathrm{ev})} M_{\mathrm{ph}}^{\mathrm{suc}} \Pi_{AC}^{(-,\mathrm{od}),(+,\mathrm{ev})} \tag{4.307}$$

$$= w_+ |-\rangle\langle-| \otimes M_{\mathrm{oo}}^{\mathrm{suc}} + w_- |+\rangle\langle+| \otimes M_{\mathrm{ee}}^{\mathrm{suc}} - w_z \left(|-\rangle\langle+| \otimes M_{\mathrm{oe}}^{\mathrm{suc}} + |+\rangle\langle-| \otimes M_{\mathrm{eo}}^{\mathrm{suc}}\right), \tag{4.308}$$

$$|\psi\rangle = \sqrt{\kappa}\, |\phi_{\mathrm{cor}}\rangle_{AC}, \tag{4.309}$$

$$\alpha = 1, \quad \gamma_+ = \gamma, \quad \gamma_- = 0, \tag{4.310}$$

where $|\phi_{\mathrm{cor}}\rangle_{AC}$ is defined in Eq. (4.191). Then, we observe

$$\lambda_{\sup}\left(\Pi_{AC}^{(-,\mathrm{od}),(+,\mathrm{ev})} M'[\kappa, \gamma] \Pi_{AC}^{(-,\mathrm{od}),(+,\mathrm{ev})}\right) \leq \lambda_{\sup}(M_{6\mathrm{d}}^{(1)}). \tag{4.311}$$

Combining inequalities (4.305) and (4.311) completes the proof. □

4.5.5 Numerical Simulations

We compute (the lower bound of) the net key gain per pulse (i.e., key rate $\check{G}$) against the transmission distance L with various values of excess noise ξ at the channel output. The channel model is the same as that in Sect. 4.4. For simplicity, the number N_{smp} of the sampling rounds is set to be $N/100$, and the bit error correction efficiency f is to be 0.95. The acceptance probability $f_{\mathrm{suc}}(q)$ is assumed to be a step function with the threshold $q_{\mathrm{th}}(> 0)$, i.e., $\Theta(q - q_{\mathrm{th}})$. The expected amplitude of coherent state β is chosen to be $\sqrt{\eta\mu}$. We set the security parameter $\varepsilon_{\mathrm{sct}} = 2^{-50}$, and set $\varepsilon_{\mathrm{cor}} = \varepsilon_{\mathrm{sec}} = \varepsilon_{\mathrm{sct}}/2$ and $\epsilon = 2^{-s} = \varepsilon_{\mathrm{sec}}^2/4$.

We set $\check{N}^{\mathrm{suc}}$, $\check{F}$, $\check{N}^{\mathrm{test}}$, and $\check{N}^{\mathrm{trash}}$ in the same way as in Sect. 4.4. We adopted $m = 1$ and $r = 0.4120$ for the function $\Lambda_{m,r}$, which leads to $(\sup \Lambda_{m,r}, \inf \Lambda_{m,r}) = (2.824, -0.9932)$. We assume that the number of "success" sampling rounds $\check{N}_{\mathrm{smp}}^{\mathrm{suc}}$ is equal to its expectation value $(P^+ + P^-)N_{\mathrm{smp}}$, where $P^{\pm}$ are defined in Eq. (4.219). We further assume that the number of bit errors $\check{E}_{\mathrm{obs}}$ observed in the sampling rounds is equal to its expectation value $P^- N_{\mathrm{smp}}$. The upper bound $\check{e}_{\mathrm{ub}}$ on the bit error rate is thus set to be

$$\check{e}_{\mathrm{ub}} = \left(\hat{M}_{\check{N}^{\mathrm{suc}} + \check{N}_{\mathrm{smp}}^{\mathrm{suc}}, \check{N}_{\mathrm{smp}}^{\mathrm{suc}}, \varepsilon_{\mathrm{cor}}/2}(\check{E}_{\mathrm{obs}}) - \check{E}_{\mathrm{obs}}\right)/\check{N}^{\mathrm{suc}}, \tag{4.312}$$

where $\hat{M}_{N,n,\epsilon}$ is defined in Corollary 4.2.12.

Under these assumptions, remaining parameters to be determined are two coefficients (κ, γ) and four protocol parameters $(\mu, q_{\mathrm{th}}, p_{\mathrm{sig}}, p_{\mathrm{test}})$, which is the same as in Sect. 4.4. We determined (κ, γ) via a convex optimization using CVXPY 1.0.25 and $(\mu, q_{\mathrm{th}}, p_{\mathrm{sig}}, p_{\mathrm{test}})$ via the Nelder-Mead in the scipy.minimize library in Python, for each transmission distance L.

Figures 4.9 and 4.10 show the key rates against transmission distance over an optical fiber with the attenuation rate η assumed to be $10^{-0.02L}$. Figure 4.9 shows that under the condition of low excess noise ($\xi = 0$–$10^{-4.0}$), our refined analysis offers

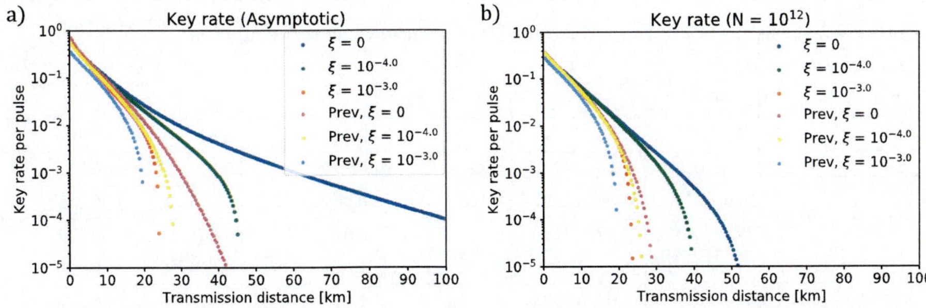

Fig. 4.9 The net key gain per pulse $\check{G}$ (key rate) against the transmission distance L of the optical channel with the refined security analysis in this section and the previous analysis in Sect. 4.4. We assumed that the optical pulse that Bob receives is given by randomly displacing a coherent state to increase its variance by a factor of $(1 + \xi)$. **a** The asymptotic key rate for various values of ξ. In the figure, "Prev" denotes the key rate computed with the previous analysis in Sect. 4.4 while $\check{N}_{\mathrm{EC}}$ is given in Eq. (4.225). **b** The key rate for various values of ξ when the pulse number is finite ($N = 10^{12}$). "Prev" denotes that with the previous analysis

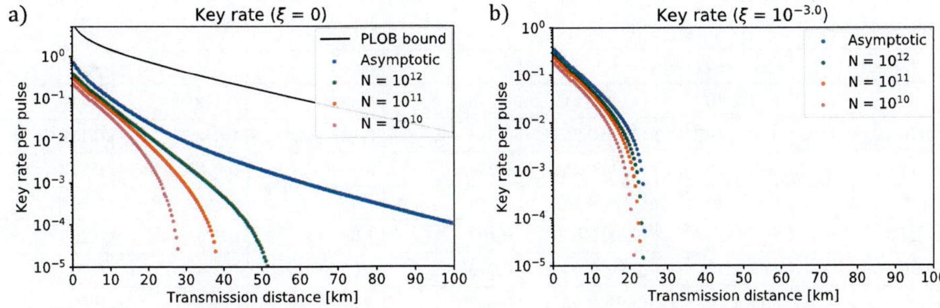

Fig. 4.10 The net key gain per pulse $\check{G}$ (key rate) against the transmission distance L of the optical channel with our refined security analysis. We assumed that the optical pulse that Bob receives is given by randomly displacing a coherent state to increase its variance by a factor of $(1 + \xi)$. **a** The key rate in a pure-loss channel ($\xi = 0$) for various pulse numbers N. The solid black line shows the PLOB bound [4]. **b** The key rate in a channel with the excess noise $\xi = 10^{-3.0}$ for various pulse numbers N

higher key rates and longer transmission distance than that of the previous section even in the finite-key case. Furthermore, as shown in Fig. 4.10a, the logarithm of the asymptotic key rate in the pure-loss case (i.e., $\xi = 0$) achieves almost linear scaling against transmission distance, which is known to be an optimal scaling allowed in the pure-loss channel. However, when the excess noise ξ is as high as $10^{-3.0}$, the improvements in our refined analysis are almost lost as shown in Figs. 4.9b and 4.10b. Compared to four-state variants of the protocol that are previously studied [40, 41, 43], our binary-modulation protocol is still fragile against excess noises.

Examples of optimized parameters are shown in Table 4.2. Overall, the key rates in Table 4.2 are better than those in Table 4.1 especially for the long transmission distances. Typical optimized values of the threshold q_{th} range from 0.7 to 1.2, which are larger than those of other analyses of protocols with post-selection (e.g., [40])

Table 4.2 Examples of optimized parameters for the reverse reconciliation

Parameters for $N = 10^{11}$ *and* $\xi = 0$

L [km]	Key rate $\check{G}$	(κ, γ)	μ	q_{th}	p_{sig}	p_{test}
5	1.25×10^{-1}	(25.8, 1.09)	0.435	0.714	0.824	0.169
10	5.38×10^{-2}	(14.6, 0.527)	0.327	0.851	0.811	0.182
15	2.22×10^{-2}	(10.4, 0.262)	0.240	0.964	0.759	0.235
20	8.48×10^{-3}	(8.07, 0.129)	0.161	1.04	0.661	0.334
25	2.99×10^{-3}	(6.25, 0.068)	0.104	1.07	0.548	0.448
30	9.07×10^{-4}	(5.03, 0.039)	0.072	1.09	0.413	0.582
35	1.77×10^{-4}	(4.22, 0.024)	0.053	1.12	0.248	0.748

Parameters for $N = 10^{11}$ *and* $\xi = 10^{-3.0}$

L [km]	Key rate $\check{G}$	(κ, γ)	μ	q_{th}	p_{sig}	p_{test}
5	1.03×10^{-1}	(14.9, 0.949)	0.389	0.750	0.849	0.142
10	4.10×10^{-2}	(9.14, 0.492)	0.305	0.890	0.830	0.162
15	1.31×10^{-2}	(6.47, 0.252)	0.233	1.03	0.765	0.226
20	1.91×10^{-3}	(4.80, 0.122)	0.164	1.15	0.561	0.428

Parameters for $N = 10^{12}$ *and* $\xi = 0$

L [km]	Key rate $\check{G}$	(κ, γ)	μ	q_{th}	p_{sig}	p_{test}
5	1.53×10^{-1}	(42.2, 1.21)	0.473	0.692	0.878	0.119
10	6.72×10^{-2}	(21.3, 0.553)	0.343	0.833	0.878	0.119
15	2.94×10^{-2}	(13.0, 0.266)	0.244	0.949	0.879	0.119
20	1.32×10^{-2}	(11.4, 0.132)	0.164	1.01	0.782	0.215
25	5.87×10^{-3}	(8.48, 0.070)	0.107	1.04	0.733	0.265
30	2.67×10^{-3}	(7.09, 0.040)	0.076	1.05	0.367	0.709
35	1.17×10^{-3}	(5.91, 0.025)	0.057	1.07	0.538	0.460
40	4.68×10^{-4}	(4.96, 0.016)	0.045	1.09	0.454	0.544
45	1.58×10^{-4}	(4.45, 0.011)	0.038	1.11	0.306	0.692
50	2.76×10^{-5}	(3.90, 0.0074)	0.032	1.13	0.165	0.834

Parameters for $N = 10^{12}$ *and* $\xi = 10^{-3.0}$

L [km]	Key rate $\check{G}$	(κ, γ)	μ	q_{th}	p_{sig}	p_{test}
5	1.20×10^{-1}	(17.9, 0.996)	0.405	0.735	0.908	0.087
10	4.94×10^{-2}	(10.3, 0.502)	0.311	0.878	0.909	0.087
15	1.73×10^{-2}	(7.09, 0.255)	0.234	1.01	0.864	0.132
20	3.48×10^{-3}	(5.03, 0.123)	0.162	1.13	0.769	0.225

Examples of the parameters for a given pair of the total pulse number N and the excess noise parameter ξ in the refined security analysis. The variance of the quadrature operator $\hat{q}$ for the vacuum state is $\langle(\Delta\hat{q})^2\rangle = 1/2$. Given (N, ξ), protocol parameters $(\kappa, \gamma, \mu, q_{\mathrm{th}}, p_{\mathrm{sig}}, p_{\mathrm{test}})$ are optimized for each transmission distance L [km] so that the net key gain per pulse (key rate) $\check{G}$ is maximized but smaller than those in Sect. 4.4. The intensities μ of the input coherent states are smaller in Table 4.2 than those in Table 4.1 for the longer transmission distances. Taking smaller intensities for longer transmission distances may be the characteristic feature of this analysis; optimal intensities for other asymptotic analyses are around 0.1–0.4 for over 100 km transmission distances [40, 41, 43].

4.5.6 Discussion

We proposed a refined security analysis for the protocol proposed in Sect. 4.4 based on the reverse reconciliation. The motivating ideas of our refinement come from the facts that the secret key can be distilled from the states from which the entanglement cannot be distilled [49, 63, 64, 87–89] and the reverse reconciliation can increase the key rate for continuous-variable QKD protocols [34]. To exploit the ideas, we developed the procedure of "twisting" Alice's system with the isometry $V_{BA\to BA'}$ controlled by the Z basis of Bob's qubit, while the techniques derived from similar ideas have already appeared in previous works [49, 63, 64, 87, 89, 90]. Our finding is that, by using the twisting operation that minimizes the phase error probability for the pure-loss channel, the protocol has asymptotically almost optimal scaling in the key rates. This is the clear distinction from the result in Sect. 4.4; there, the logarithm of the asymptotic key rate non-linearly decreases against transmission distance. The improvement in the performance remains in the finite-key case but is lost under the existence of excess noise as high as $\xi = 10^{-3.0}$ at the channel output. This may limit the feasibility of our binary-modulation protocol. In fact, weakness against excess noises may be universal for binary-modulation continuous-variable QKD protocols [38]. Current theoretical progress in the continuous-variable QKD reveals that the discrete-modulation continuous-variable QKD protocols with four types of modulation have more tolerance against excess noise than those with binary modulation [40, 41, 43]. What is important is that our security proof can be extended to the four-state protocols with binary outcomes, such as Protocol 2 in Ref. [40] and a protocol in Ref. [43], by replacing the bit-extracting measurements of these protocols with the qubit-extracting maps as shown in Eq. (4.118) and defining the corresponding phase error operator. This is, however, much more complicated than the previous analysis, and we leave the problem as future work.

About the practicality of the protocol, reducing the total number 10^{12} of rounds may be desired to save the cost of classical information processing. The finite-size performance of our protocol may be improved by applying recently developed refinement [59] of the Azuma's inequality [56] that utilizes unconfirmed knowledge. What is non-trivial for applying this is that the random variable in our application of Azuma's inequality can not directly be observed even at the end of the protocol. Whether we can obtain a tighter bound using the refined concentration inequality [59] with the information accessible in our protocol is an open problem.

4.6 Conclusion for This Chapter

In this chapter, the finite-size security of the binary-modulation continuous-variable QKD protocol is established. The key to our achievement is the newly developed fidelity estimation for the channel output using the heterodyne measurement in Sect. 4.4.1. The fidelity as a measure of disturbance is essentially the same as what is monitored through bit errors in the B92 protocol [6, 71, 72], and thus we can apply the proof techniques of the discrete-variable QKDs. This reduction has, however, non-triviality in the operator inequality (4.140) or (4.255) due to its infinite dimensionality. This is circumvented in Sects. 4.4.4 or 4.5.4 by constructing the heuristic but fairly good upper bound that is essentially finite-dimensional on the considered operator. These analyses are adapted to the digitization of the classical information processing; the detectors with finite resolution can be treated in our security analysis with minimal degradation to the protocol performance. In this sense, our security analysis is for digital information processing while the information itself is carried by the continuous-variable system.

While it is simple, the entanglement-distillation-based security proof developed in Sect. 4.4 turns out not to be tight. A tighter analysis with improved performance for the same protocol can be obtained via the reverse reconciliation in Sect. 4.5 at the cost of an additional complication for the operator inequality (Sect. 4.5.4). This refined analysis gives an almost optimal scaling of the asymptotic key rate against transmission distance. Our refined analysis is still fragile against the excess noise; Fig. 4.10 b shows that the improvement of the performance with our refined analysis is lost by the excess noise as high as $\xi = 10^{-3.0}$ at the channel output. The key rates under the higher excess noise shown in Fig. 4.7 are not practical. This may be a fate of the binary modulation protocol; the similar protocol considered in Ref. [38] shows similar behaviors under the existence of the excess noise.

To improve the performance under the existence of the excess noise, therefore, we need to extend our finite-size analysis to protocols with more constellations. Among other things, the four-state protocols have been shown to be more robust against the excess noise [41, 41, 43]. Our analysis can, in principle, be extended to the four-state protocols with binary outputs [40, 43], i.e., protocols that use homodyne measurement to distinguish signals, by replacing the bit extraction with the qubit extraction as shown in Eq. (4.118). However, developing the operator inequalities that keep the robustness against excess noise of these protocols still has non-triviality. A more challenging problem is to apply our finite-size security proof to the four-state protocols with more than two outputs, such as a protocol in Ref. [41] and Protocol 1 in Ref. [40]. In this case, the definition of phase errors is already non-trivial as opposed to those with binary outputs, and we have to develop more elaborated finite-size security proof. Whether we can extend our techniques to these protocols or protocols with even more constellations is still open. The results in this chapter can thus be regarded as a first step for the complete security proofs for the continuous-variable QKD with digitized information processing.

References

1. Bennett CH, Brassard G (1984) Quantum cryptography: public key distribution and coin tossing. In: Proceedings of IEEE international conference on computers, systems, and signal processing. Bangalore, India, p 175
2. Gottesman D, Kitaev A, Preskill J (2001) Encoding a qubit in an oscillator. Phys Rev 64(1):012310
3. Tomamichel M (2012) A Framework for Non-Asymptotic Quantum Information Theory. PhD thesis. ETH Zurich
4. Pirandola S et al (2017) Fundamental limits of repeaterless quantum communications. Nat Commun 8(1):15043. issn: 2041-1723. https://doi.org/10.1038/ncomms15043
5. Shannon CE (1949) Communication theory of secrecy systems. Bell Syst Tech J 28(4):656–715
6. Bennett CH (1992) Quantum cryptography using any two nonorthogonal states. Phys Rev Lett 68(21):3121
7. Ralph TC (1999) Continuous variable quantum cryptography. Phys Rev 61(1):010303
8. Hillery M (2000) Quantum cryptography with squeezed states. Phys Rev 61(2):022309
9. Cerf NJ, Lévy M, Van Assche G (2001) Quantum distribution of Gaussian keys using squeezed states. Phys Rev 63:052311. https://doi.org/10.1103/PhysRevA.63.052311
10. Grosshans F, Grangier P (2002) Continuous variable quantum cryptography using coherent states. Phys Rev Lett 88(5):057902
11. Grosshans F et al (2003) Quantum key distribution using gaussian-modulated coherent states. Nature 421(6920):238–241
12. Weedbrook C et al (2004) Quantum cryptography without switching. Phys Rev Lett 93(17):170504
13. Xu F et al (2020) Secure quantum key distribution with realistic devices. Rev Mod Phys 92:025002. https://doi.org/10.1103/RevModPhys.92.025002
14. Pirandola S et al (2020) Advances in quantum cryptography. Adv Opt Photon 12(4):1012–1236. https://doi.org/10.1364/AOP.361502, http://www.osapublishing.org/aop/abstract.cfm?URI=aop-12-4-1012
15. Eriksson TA et al (2020) Wavelength division multiplexing of 194 continuous variable quantum key distribution channels. J Light Technol 38(8):2214–2218. https://doi.org/10.1109/JLT.2020.2970179
16. Huang D et al (2015) Continuous-variable quantum key distribution with 1 Mbps secure key rate. Opt Express 23(13):17511–17519
17. Kumar R, Qin H, Alléaume R (2015) Coexistence of continuous variable QKD with intense DWDM classical channels. New J Phys 17(4):043027
18. Huang D et al (2016) Field demonstration of a continuous-variable quantum key distribution network. Opt Lett 41(15):3511–3514
19. Karinou F et al (2017) Experimental evaluation of the impairments on a QKD system in a 20-channel WDM co-existence scheme. In: 2017 IEEE photonics society summer topical meeting series (SUM). IEEE, pp 145–146
20. Karinou F et al (2018) Toward the integration of CV quantum key distribution in deployed optical networks. IEEE Photonics Technol Lett 30(7):650–653
21. Eriksson TA et al (2018) Coexistence of continuous variable quantum key distribution and 7 × 12.5 Gbit, s classical channels. In: 2018 IEEE photonics society summer topical meeting series (SUM). IEEE, pp 71–72
22. Eriksson TA et al (2019) Wavelength division multiplexing of continuous variable quantum key distribution and 18.3 tbit/s data channels. Commun Phys 2(1):1–8
23. Milovancev D et al (2020) Spectrally-shaped continuous-variable QKD operating at 500 MHz over an optical pipe lit by 11 DWDM channels. In: 2020 optical fiber communications conference and exhibition (OFC), pp 1– 3
24. Diamanti E, Leverrier A (2015) Distributing secret keys with quantum continuous variables: principle, security and implementations. Entropy 17(9):6072–6092

25. Laudenbach F et al (2018) Continuous-variable quantum key distribution with gaussian modulation-the theory of practical implementations. Adv Quantum Technol 1(1):1800011
26. Leverrier A et al (2009) Security of continuous-variable quantum key distribution: towards a de Finetti theorem for rotation symmetry in phase space. New J Phys 11(11):115009. https://doi.org/10.1088/1367-2630/11/11/115009
27. Leverrier A (2017) SU(q, p) coherent states and a Gaussian de Finetti theorem. arXiv: 1612.05080 [quant-ph]
28. Leverrier A et al (2013) Security of continuous-variable quantum key distribution against general attacks. Phys Rev Lett 110:030502. https://doi.org/10.1103/PhysRevLett.110.030502
29. Leverrier A (2015) Composable security proof for continuous-variable quantum key distribution with coherent states. Phys Rev Lett 114:070501. https://doi.org/10.1103/PhysRevLett.114.070501
30. Leverrier A (2017) Security of continuous-variable quantum key distribution via a Gaussian de Finetti reduction. Phys Rev Lett 118(20):200501
31. Jouguet P et al (2012) Analysis of imperfections in practical continuous-variable quantum key distribution. Phys Rev 86(3):032309
32. Lupo C (2020) Towards practical security of continuous-variable quantum key distribution. Phys Rev 102:022623. https://doi.org/10.1103/PhysRevA.102.022623
33. Kaur E, Guha S, Wilde MM (2021) Asymptotic security of discretemodulation protocols for continuous-variable quantum key distribution. Phys Rev 103:012412. https://doi.org/10.1103/PhysRevA.103.012412
34. Silberhorn C et al (2002) Continuous variable quantum cryptography: Beating the 3 dB loss limit. Phys Rev Lett 89(16):167901
35. Hirano T et al (2003) Quantum cryptography using pulsed homodyne detection. Phys Rev 68(4):042331
36. Leverrier A, Grangier P (2009) Unconditional security proof of longdistance continuous-variable quantum key distribution with discrete modulation. Phys Rev Lett 102(18):180504
37. Leverrier A, Grangier P (2011) Erratum: unconditional security proof of long-distance continuous-variable quantum key distribution with discrete modulation [Phys Rev Lett 102:180504 (2009)]. Phys Rev Lett 106:259902. https://doi.org/10.1103/PhysRevLett.106.259902
38. Zhao Y-B et al (2009) Asymptotic security of binary modulated continuous-variable quantum key distribution under collective attacks. Phys Rev 79(1):012307
39. Brádler K, Weedbrook C (2018) Security proof of continuous-variable quantum key distribution using three coherent states. Phys Rev 97(2):022310
40. Lin J, Upadhyaya T, Lütkenhaus N (2019) Asymptotic security analysis of discrete-modulated continuous-variable quantum key distribution. Phys Rev 9(4):041064
41. Ghorai S et al (2019) Asymptotic security of continuous-variable quantum key distribution with a discrete modulation. Phys Rev 9(2):021059
42. Denys A, Brown P, Leverrier A (2021) Explicit asymptotic secret key rate of continuous-variable quantum key distribution with an arbitrary modulation. Quantum 5:540. issn: 2521-327X. https://doi.org/10.22331/q-2021-09-13-540
43. Liu W-B et al (2021) Homodyne detection quadrature phase shift keying continuous- variable quantum key distribution with high excess noise tolerance. PRX Quantum 2:040334. https://doi.org/10.1103/PRXQuantum.2.040334
44. Papanastasiou P, Pirandola S (2021) Continuous-variable quantum cryptography with discrete alphabets: composable security under collective Gaussian attacks. Phys Rev Res 3:013047. https://doi.org/10.1103/PhysRevResearch.3.013047
45. Papanastasiou P, Ottaviani C, Pirandola S (2021) Security of continuous-variable quantum key distribution against canonical attacks. 2021 international conference on computer communications and networks (ICCCN), pp 1–6. https://doi.org/10.1109/ICCCN52240.2021.9522349
46. Matsuura T et al (2021) Finite-size security of continuous-variable quantum key distribution with digital signal processing. Nat Commun 12(1):252. issn: 2041-1723. https://doi.org/10.1038/s41467-020-19916-1

47. Cover TM, Thomas JA (2012) Elements of information theory. Wiley, New York (2012)
48. Renner R, König R (2005) Universally composable privacy amplification against quantum adversaries. In: Kilian J (ed) Theory of cryptography. Springer, Berlin, pp 407–425. isbn: 978-3-540-30576-7
49. Renner R (2008) Security of quantum key distribution. Int J Quantum Inf 6(01):1–127
50. Konig R, Renner R, Schaffner C (2009) The operational meaning of min- and max-entropy. IEEE Trans Inf Theory 55(9):4337–4347. https://doi.org/10.1109/TIT.2009.2025545
51. Tomamichel M, Colbeck R, Renner R (2009) A fully quantum asymptotic equipartition property. IEEE Trans Inf Theory 55(12):5840–5847. https://doi.org/10.1109/TIT.2009.2032797
52. Chernoff H (1952) A measure of asymptotic efficiency for tests of a hypothesis based on the sum of observations. Ann Math Stat 23(4):493–507. https://doi.org/10.1214/aoms/1177729330
53. Hoeffding W (1963) Probability inequalities for sums of bounded random variables. J Am Stat Assoc 58(301):13–30. https://doi.org/10.1080/01621459.1963.10500830
54. Hoeffding W (1994) Probability inequalities for sums of bounded random variables. In: The collected works of Wassily Hoeffding. Springer, Berlin, pp 409–426
55. Chvátal V (1979) The tail of the hypergeometric distribution. Discret Math 25(3):285–287. issn: 0012-365X. https://doi.org/10.1016/0012-365X(79)90084-0
56. Azuma K (1967) Weighted sums of certain dependent random variables. Tohoku Math J Second Ser 19(3):357–367
57. Raginsky M, Sason I (2013) Concentration of measure inequalities in information theory, communications, and coding. Found Trends Commun Inf Theory 10(1–2):1–246. https://doi.org/10.1561/0100000064
58. McDiarmid C (1998) Concentration. In: Probabilistic methods for algorithmic discrete mathematics. Springer, Berlin, pp 195–248
59. Kato G (2020) Concentration inequality using unconfirmed knowledge. arXiv: 2002.04357 [math.PR]
60. Müller-Quade J, Renner R (2009) Composability in quantum cryptography. New J Phys 11(8):085006
61. Lo HK, Chau HF (1999) Unconditional security of quantum key distribution over arbitrarily long distances. Science 283(5410):2050–2056
62. Shor PW, Preskill J (2000) Simple proof of security of the BB84 quantum key distribution protocol. Phys Rev Lett 85(2):441
63. Horodecki K et al (2009) General paradigm for distilling classical key from quantum states. IEEE Trans Inf Theory 55(4):1898–1929. https://doi.org/10.1109/TIT.2008.2009798
64. Koashi M (2009) Simple security proof of quantum key distribution based on complementarity. New J Phys 11(4):045018
65. Tsurumaru T (2020) Leftover hashing from quantum error correction: unifying the two approaches to the security proof of quantum key distribution. IEEE Trans Inf Theory 66(6):3465–3484. https://doi.org/10.1109/TIT.2020.2969656
66. Tsurumaru T (2020) Equivalence of three quantum algorithms: privacy amplification, error correction, and data compression. arXiv: 2009.08823 [quant-ph]
67. Carter JL, Wegman MN (1979) Universal classes of hash functions. J Comput Syst Sci 18(2):143–154
68. Tsurumaru T, Hayashi M (2013) Dual universality of hash functions and its applications to quantum cryptography. IEEE Trans Inf Theory 59(7):4700–4717. https://doi.org/10.1109/TIT.2013.2250576
69. Matsuura T, Sasaki T, Koashi M (2019) Refined security proof of the round-robin differential-phase-shift quantum key distribution and its improved performance in the finite-sized case. Phys Rev 99(4):042303
70. Lupo C, Ouyang Y (2021) Quantum key distribution with non-ideal heterodyne detection: composable security of discrete-modulation continuousvariable protocols. arXiv: 2108.00428 [quant-ph]

71. Tamaki K, Koashi M, Imoto N (2003) Unconditionally secure key distribution based on two nonorthogonal states. Phys Rev Lett 90(16):167904
72. Koashi M (2004) Unconditional security of coherent-state quantum key distribution with a strong phase-reference pulse. Phys Rev Lett 93(12):120501
73. Shor PW, Preskill J (2000) Simple proof of security of the BB84 quantum key distribution protocol. Phys Rev Lett 85(2):441
74. Abramowitz M, Stegun IA (1948) Handbook of mathematical functions with formulas, graphs, and mathematical tables, vol 55. US Government printing office
75. Chabaud U et al (1905) Building trust for continuous variable quantum states. 12700:2019
76. Chabaud U et al (2021) efficient verification of boson sampling. Quantum 5:578. issn: 2521-327X. https://doi.org/10.22331/q-2021-11-15-578
77. Hayashi M, Tsurumaru T (2012) Concise and tight security analysis of the Bennett-Brassard 1984 protocol with finite key lengths. New J Phys 14(9):093014
78. Namiki R, Hirano T (2004) Practical limitation for continuous-variable quantum cryptography using coherent states. Phys Rev Lett 92(11):117901
79. Hirano T et al (2017) Implementation of continuous-variable quantum key distribution with discrete modulation. Quantum Sci Technol 2(2):024010
80. Diamond S, Boyd S (2016) CVXPY: a Python-embedded modeling language for convex optimization. J Mach Learn Res 17(83):1–5
81. Agrawal A et al (2018) A rewriting system for convex optimization problems. J Control Decis 5(1):42–60
82. Jouguet P et al (2013) Experimental demonstration of long-distance continuousvariable quantum key distribution. Nat Photonics 7(5):378–381
83. Devetak I, Winter A (2005) Distillation of secret key and entanglement from quantum states. In: Proc R Soc A: Math, Phys Eng Sci 461(2053):207–235
84. Lodewyck J et al (2007) Quantum key distribution over 25 km with an allfiber continuous-variable system. Phys Rev 76:042305. https://doi.org/10.1103/PhysRevA.76.042305
85. Jouguet P, Elkouss D, Kunz-Jacques S (2014) High-bit-rate continuousvariable quantum key distribution. Phys Rev 90:042329. https://doi.org/10.1103/PhysRevA.90.042329
86. Wen X et al (2021) Rotation based slice error correction protocol for continuousvariable quantum key distribution and its implementation with polar codes. arXiv: 2106.06206 [quant-ph]
87. Horodecki K et al (2006) Quantum key distribution based on arbitrarily weak distillable entangled states. Phys Rev Lett 96:070501. https://doi.org/10.1103/PhysRevLett.96.070501
88. Renes JM, Smith G (2007) Noisy processing and distillation of private quantum states. Phys Rev Lett 98:020502. https://doi.org/10.1103/PhysRevLett.98.020502
89. Horodecki K et al (2008) Unconditional privacy over channels which cannot convey quantum information. Phys Rev Lett 100:110502. https://doi.org/10.1103/PhysRevLett.100.110502
90. Eli Bourassa J et al (2020) Loss-tolerant quantum key distribution with mixed signal states. Phys Rev 102:062607. https://doi.org/10.1103/PhysRevA.102.062607

Chapter 5
Quantum Computation with Continuous-Variable Systems

Abstract In this chapter, the Gottesman-Kitaev-Preskill (GKP) code, a quantum error-correcting code for continuous-variable quantum information processing, is studied in the context of optical quantum computation. Especially, the equivalence of its approximations is shown, and the cost-efficient implementation of the universal optical quantum computation using the "GKP magic state" is developed.

Keywords Optical quantum computation · The Gottesman-Kitaev-Preskill code · Universality · Magic state

5.1 Introduction for This Chapter

Quantum computation (QC) brings advantages over conventional classical computation in terms of computational speedups [5–7] and stronger security [8, 9]. Continuous-variable systems, especially quantum optical systems, have attracted growing interest as promising candidates for implementing QC. Compared to other matter-based candidates for implementing QC such as superconducting qubits [10, 11] and ion traps [12, 13], characteristics of the quantum optical architectures are scalability in generating quantum entanglement among more than one million photonic modes [14] and flexibility in interacting photons that fly in space, being free from geometrical constraints of two-dimensional surfaces of the matters.

For fault-tolerant quantum computing, one needs to construct an error-correcting routine to fight against the inevitable noise in the real world. Intensive research has thus been made on continuous-variable error-correcting codes [1, 15–31]. Among them, the Gottesman-Kitaev-Preskill (GKP) code [1], which encodes a qudit into an oscillator, gathers much attention in terms of both fault-tolerant continuous-variable QC [17, 32–40] and continuous-variable quantum communication [31, 41, 42]. It needs only Gaussian operations, which is tractable in the quantum optical experiments, to implement Clifford gates (or even the universal gate set using protocols with a single type of GKP-encoded state [43]), and it is highly robust against loss errors as well as random displacement errors [33, 42].

T. Matsuura, *Digital Quantum Information Processing with Continuous-Variable Systems*, Springer Theses, https://doi.org/10.1007/978-981-19-8288-0_5

In this chapter, we first show the equivalence between different conventional approximations of the GKP code. (The reason why approximations are needed will be explained in detail later.) This bridges the gap of previous studies that were based on different approximations. We then develop the strategy of efficiently implementing the universal QC with the GKP code and Gaussian operations, which is suited for optical systems. The chapter is organized as follows. Section 5.2.2 summarizes the preliminaries about the GKP code. Sections 5.3 and 5.4 are the results of this thesis. In Sect. 5.3, the equivalence between the conventional approximations of the GKP code is proved, and the standard form of the approximate GKP codes is introduced. Furthermore, the explicit forms of the normalization constants and the average photon numbers of the approximate GKP code are derived in terms of the standard form. In Sect. 5.4, a resource-efficient implementation of the universal QC with a single type of GKP-encoded state and the Gaussian operations is constructed. A comparison between our construction and the previous proposal is discussed from a resource-theoretic perspective. Especially, the degree of non-Gaussianity of the approximate GKP-encoded state is computed numerically using the standard form developed in Sect. 5.3. Based on the existing proposals of preparing GKP-encoded states, we further discuss the feasibility of our constructed protocol (Sects. 5.3 and 5.4 are based on the publications [44] and [45][1].).

5.2 Notations and Preliminaries

5.2.1 *Qudit, The Pauli Group, and The Clifford Group*

Qudit is a quantum system characterized by the d-dimensional Hilbert space. It is the generalization of the qubit. Let $\{|j\rangle : j = 0, \ldots, d-1\}$ be an orthonormal basis of a qubit. Then, we can define the (generalized) Pauli-Z and -X operators by

$$Z|j\rangle = \exp(2\pi i j/d)|j\rangle, \qquad X|j\rangle = |j+1\rangle, \tag{5.1}$$

where the summation is modulo d. The commutation relation between Z and X are thus given by $ZX = \exp(2\pi i/d)XZ$. The group generated by Z, X, and iI is called the (generalized) Pauli group. Note that the global phase factor iI is irrelevant in quantum theory.

In QC, the normalizer of the Pauli group called the Clifford group plays an important role. In the case of the qubit, the Clifford group is generated by the Hadamard gate H, the phase gate S, and the CNOT gate given respectively by

$$H = \frac{1}{\sqrt{2}} \begin{pmatrix} 1 & 1 \\ 1 & -1 \end{pmatrix}, \tag{5.2}$$

$$S = \begin{pmatrix} 1 & 0 \\ 0 & i \end{pmatrix}, \tag{5.3}$$

$$\text{CNOT} = |0\rangle\langle 0|_1 \otimes I_2 + |1\rangle\langle 1|_1 \otimes \sigma_2^X. \tag{5.4}$$

These gates transform the Pauli-Z and -X operators by the adjoint actions as follows:

$$H^\dagger \sigma^Z H = \sigma^X, \qquad H^\dagger \sigma^X H = \sigma^Z, \tag{5.5}$$

$$S^\dagger \sigma^Z S = \sigma^Z, \qquad S^\dagger \sigma^X S = -\sigma^Y = i\sigma^Z \sigma^X, \tag{5.6}$$

$$\text{CNOT}^\dagger (\sigma_1^Z \otimes I_2) \text{CNOT} = \sigma_1^Z \otimes I_2, \qquad \text{CNOT}^\dagger (\sigma_1^X \otimes I_2) \text{CNOT} = \sigma_1^X \otimes \sigma_2^X, \tag{5.7}$$

$$\text{CNOT}^\dagger (I_1 \otimes \sigma_2^Z) \text{CNOT} = \sigma_1^Z \otimes \sigma_2^Z, \qquad \text{CNOT}^\dagger (I_1 \otimes \sigma_2^X) \text{CNOT} = I_1 \otimes \sigma_2^X. \tag{5.8}$$

It is known that all the unitaries on an N-qudit system can be approximately generated by the combination of Clifford gates and a non-Clifford gate in an arbitrary accuracy [2, 46, 47], while cannot be by only Clifford gates [48, 49]. The set of gates that can approximately generate an arbitrary N-qudit unitary is called the universal gate set.

5.2.2 The Gottesman-Kitaev-Preskill Code

In this chapter, we add a subscript of the basis choice instead of the system to a state vector for clarity, e.g., $|\psi\rangle_f$ for the Fock basis and $|q\rangle_{\hat{q}}$ for the quadrature "basis" in the sense of Eq. (3.9). (Strictly speaking, $|\cdot\rangle_{\hat{q}}$ is not an element of the Hilbert space for the oscillator mode, so it should be regarded as a weak limit of the squeezed coherent state as the squeezing parameter, i.e., ξ in Eq. (3.91), goes to infinity.) Exceptionally, the state vector with no subscripts denotes the GKP-encoded state. The GKP code [1] is an error-correcting code that encodes a qudit into an oscillator mode. It has a lattice-like periodic structure when represented in the phase space; the Wigner function of the code states $|j\rangle$ and $|j+1\rangle$ have the same period, but $|j+1\rangle$ is shifted from $|j\rangle$ by $\frac{1}{d}$ of the period in position, where $j \in \{0, \ldots, d-1\}$. The ideal (square lattice) GKP logical basis states $\{|j^{(\text{ideal})}\rangle : j = 0, \ldots, d-1\}$ are defined as [1].

$$|j^{(\text{ideal})}\rangle := \sqrt{\alpha d} \sum_{s \in \mathbb{Z}} |\alpha(ds + j)\rangle_{\hat{q}}, \tag{5.9}$$

where $|\cdot\rangle_{\hat{q}}$ denotes the position "basis", and the pre-factor $\sqrt{\alpha d}$ is for later convenience. It is not actually an element of the Hilbert space; how to treat it is explained later. In the position wave function representation, it has a comb-like shape consisting of the Dirac delta functions (i.e., a Dirac comb) at intervals αd, and $|j+1^{(\mathrm{ideal})}\rangle$ is shifted from $|j^{(\mathrm{ideal})}\rangle$ by α. In the momentum representation, the logical basis states are given by

$$|j^{(\mathrm{ideal})}\rangle = \int dy \sqrt{\alpha d} \sum_{s\in\mathbb{Z}} |y\rangle\langle y|_{\hat{p}} |\alpha(ds+j)\rangle_{\hat{q}} \tag{5.10}$$

$$= \sqrt{\frac{\alpha d}{2\pi}} \int dy \sum_{s\in\mathbb{Z}} e^{-i\alpha(ds+j)p} |y\rangle_{\hat{p}} \tag{5.11}$$

$$= \sqrt{2\pi\alpha d} \int dy \sum_{t\in\mathbb{Z}} \delta(\alpha d y - 2\pi t) e^{-ij\alpha p} |y\rangle_{\hat{p}} \tag{5.12}$$

$$= \sqrt{\frac{2\pi}{\alpha d}} \sum_{t\in\mathbb{Z}} e^{-i\frac{2\pi j t}{d}} |2\pi t/(\alpha d)\rangle_{\hat{p}}, \tag{5.13}$$

where we used the Poisson summation formula $\sum_{s\in\mathbb{Z}} e^{-isx} = 2\pi \sum_{t\in\mathbb{Z}} \delta(x-2\pi t)$. The parameter α is often chosen to be

$$\alpha = \sqrt{\frac{2\pi}{d}} =: \alpha_d, \tag{5.14}$$

in order to symmetrize the code space in position and momentum coordinates in the phase space [1], where the "symmetric code" is defined as follows.

Definition 5.2.1 (*The code that is symmetric in position and momentum coordinates in the phase space*) Let $\{|j\rangle : j = 0, \ldots, d-1\}$ be the logical qudit basis encoded in an oscillator mode. The code is symmetric in position and momentum coordinates if it satisfies

$$\mathrm{span}\{|j\rangle : j = 0, \ldots, d-1\} = \mathrm{span}\{R(\pi/2)|j\rangle : j = 0, \ldots, d-1\}, \tag{5.15}$$

where $R(\pi/2)$ denotes the Fourier transform defined in Eq. (3.83).

Note that we can use $R(-\pi/2)$ instead of $R(\pi/2)$ in the definition. We also adopt Eq. (5.14) in the following. The symmetric code is beneficial if we aim at not biasing logical-level errors caused by physical-level phase-insensitive errors, that is, errors that occur symmetrically in position and momentum coordinates in the phase space. Furthermore, this definition implies that the Fourier transform $R(\pi/2)$ is an element of the stabilizer or a logical operator of the code since it preserves the code space. This symmetrization of the code is meaningful even when the logical basis states are nonorthogonal, which is the case in approximate GKP codes.

The ideal GKP code can be regarded as a stabilizer code. The stabilizer generators are given by the two commuting displacement operators $X_{\mathrm{st}} := X(\alpha_d d)$ and $Z_{\mathrm{st}} := Z(2\pi/\alpha_d) = Z(\alpha_d d)$, where $X(s) := \exp(-is\hat{p})$ and $Z(t) := \exp(it\hat{q})$ are introduced in Eq. (3.2). The logical Pauli-X and Z operators can be defined as $X_L := X(\alpha_d)$ and $Z_L := Z(2\pi/(\alpha_d d)) = Z(\alpha_d)$, which satisfy $Z_L X_L = \exp(2\pi i/d) X_L Z_L$ as expected. The GKP-encoded states defined in Eq. (5.9) are stabilized by $X_{\rm st}$ and $Z_{\rm st}$, and are the eigenstates of the logical operator Z_L. The logical Clifford operators on the GKP code can be realized by the (symplectic) linear transformation of the quadrature operators $\hat{q}$ and $\hat{p}$ since logical Pauli operators are of the form $\exp[i(a\hat{q}+b\hat{p})]$ $(a, b \in \mathbb{R})$ up to a global phase. As explained in Sect. 3.1, the symplectic linear transformation of the quadrature gates can be implemented by Gaussian unitaries. Therefore, the logical Clifford gates on the GKP-encoded states can be realized by the Gaussian unitaries, which are reliably implementable in the quantum optical system. This is one of the advantages of using the GKP code in quantum optical systems.

For the GKP code with $d = 2$, the Gaussian unitaries that correspond to the logical H, S, and CNOT gates are given respectively by [1]

$$H \longrightarrow \exp(\pi i\hat{n}/2)(= R(\pi/2)), \tag{5.16}$$

$$S \longrightarrow \exp(i\hat{q}^2/2), \tag{5.17}$$

$$\text{CNOT} \longrightarrow \exp(-i\hat{q}_1\hat{p}_2). \tag{5.18}$$

These relations are understood by noticing that the following transformation rules hold:

$$R(-\pi/2)Z_L R(\pi/2) = X_L, \quad R(-\pi/2)X_L R(\pi/2) = Z(-\alpha_d) = Z_L^\dagger = Z_L Z_{\rm st}^\dagger, \tag{5.19}$$

$$e^{-i\hat{q}^2/2} Z_L e^{i\hat{q}^2/2} = Z_L, \quad e^{-i\hat{q}^2/2} X_L e^{i\hat{q}^2/2} = e^{-i\alpha_d\hat{q}-i\alpha_d\hat{p}} = iZ_L X_L^\dagger = iZ_L X_L X_{\rm st}^\dagger, \tag{5.20}$$

$$e^{i\hat{q}_1\hat{p}_2}(Z_L\otimes I_L)e^{-i\hat{q}_1\hat{p}_2} = Z_L\otimes I_L, \quad e^{i\hat{q}_1\hat{p}_2}(X_L\otimes I_L)e^{-i\hat{q}_1\hat{p}_2} = X_L\otimes X_L^\dagger = X_L\otimes X_L X_{\rm st}^\dagger, \tag{5.21}$$

$$e^{i\hat{q}_1\hat{p}_2}(I_L\otimes Z_L)e^{-i\hat{q}_1\hat{p}_2} = Z_L\otimes Z_L, \quad e^{i\hat{q}_1\hat{p}_2}(I_L\otimes X_L)e^{-i\hat{q}_1\hat{p}_2} = I_L\otimes X_L, \tag{5.22}$$

where we used Eqs. (3.6) and (3.83). We see that they are equivalent to Eqs. (5.5)–(5.8) up to the stabilizer elements $X_{\rm st}$ and $Z_{\rm st}$. The logical Z-basis measurement for the GKP code should distinguish $|0^{\rm (ideal)}\rangle$ and $|1^{\rm (ideal)}\rangle$ state. This can be achieved by performing the q-homodyne measurement, which always outputs an integer multiple of $\sqrt{\pi}$ for the GKP-encoded state. If the outcome is an even(odd) integer multiple of $\sqrt{\pi}$, then we know that $|0(1)^{\rm (ideal)}\rangle$ is observed. In fact, all the Pauli measurements can be realized by the homodyne measurements with appropriate directions in the phase space.

Using the stabilizer generators and logical Pauli operators, we have an alternative expression of the ideal GKP logical state as follows [1, 31]:

$$|j^{(\mathrm{ideal})}\rangle = \frac{(2d)^{-\frac{1}{4}}}{\vartheta(0, id)} \sum_{\substack{s_1 \in \mathbb{Z} \\ s_2 \in \mathbb{Z}}} X(\alpha_d(ds_1 + j)) Z(\alpha_d s_2) |0\rangle_f \tag{5.23}$$

$$= \frac{(2d)^{-\frac{1}{4}}}{\vartheta(0, id)} X_L{}^j \left(\sum_{l=0}^{d-1} Z_L{}^l \right) \sum_{\substack{s_1 \in \mathbb{Z} \\ s_2 \in \mathbb{Z}}} X_{\mathrm{st}}{}^{s_1} Z_{\mathrm{st}}{}^{s_2} |0\rangle_f \tag{5.24}$$

$$=: X_L{}^j \left(\sum_{l=0}^{d-1} Z_L{}^l \right) P_{\mathrm{GKP}} |0\rangle_f, \tag{5.25}$$

where $\vartheta(0, id) = \sum_{s \in \mathbb{Z}} \exp(-\pi d s^2)$ is the theta function, which will be explained later, and the last line defines an operator P_{GKP}, which is interpreted as the projection onto the code space ignoring the normalization. The consistency with Eq. (5.9) can be confirmed as follows [31]:

$$\frac{(2d)^{-\frac{1}{4}}}{\vartheta(0, id)} \sum_{\substack{s_1 \in \mathbb{Z} \\ s_2 \in \mathbb{Z}}} X(\alpha_d(ds_1 + j)) Z(\alpha_d s_2) |0\rangle_f$$

$$= \frac{(2d)^{-\frac{1}{4}}}{\vartheta(0, id)} \sum_{\substack{s_1 \in \mathbb{Z} \\ s_2 \in \mathbb{Z}}} \int dq\, e^{iq\alpha_d s_2} |q + \alpha_d(ds_1 + j)\rangle\langle q|_{\hat{q}} |0\rangle_f \tag{5.26}$$

$$= \frac{(2\pi d)^{-\frac{1}{4}}}{\vartheta(0, id)} \sum_{\substack{s_1 \in \mathbb{Z} \\ s_2 \in \mathbb{Z}}} \int dq\, e^{-\frac{1}{2}q^2 + iq\alpha_d s_2} |q + \alpha_d(ds_1 + j)\rangle_{\hat{q}} \tag{5.27}$$

$$= \frac{\sqrt{2\pi\alpha_d}}{\vartheta(0, id)} \sum_{\substack{s_1 \in \mathbb{Z} \\ s_2' \in \mathbb{Z}}} \int dq\, e^{-\frac{1}{2}q^2} \delta(q\alpha_d + 2\pi s_2') |q + \alpha_d(ds_1 + j)\rangle_{\hat{q}} \tag{5.28}$$

$$= \frac{\sqrt{\alpha_d d}}{\vartheta(0, id)} \sum_{\substack{s_1 \in \mathbb{Z} \\ s_2' \in \mathbb{Z}}} e^{-\pi d s_2'^2} |q + \alpha_d\big(d(s_1 - s_2') + j\big)\rangle_{\hat{q}} \tag{5.29}$$

$$= \sqrt{\alpha_d d} \sum_{s_1' \in \mathbb{Z}} |q + \alpha_d(ds_1' + j)\rangle_{\hat{q}} \tag{5.30}$$

$$= |j^{(\mathrm{ideal})}\rangle, \tag{5.31}$$

where we used $\langle q|0\rangle_f = \pi^{-\frac{1}{4}} \exp(-q^2/2)$ in the second equality, used the Poisson summation formula $\sum_{s_2 \in \mathbb{Z}} e^{-is_2 x} = 2\pi \sum_{s_2' \in \mathbb{Z}} \delta(x - 2\pi s_2')$ in the third equality, and defined $s_1' := s_1 - s_2'$ in the fifth equality.

In the phase space, the Wigner function of the state $|j^{(\mathrm{ideal})}\rangle$ is given by [1]

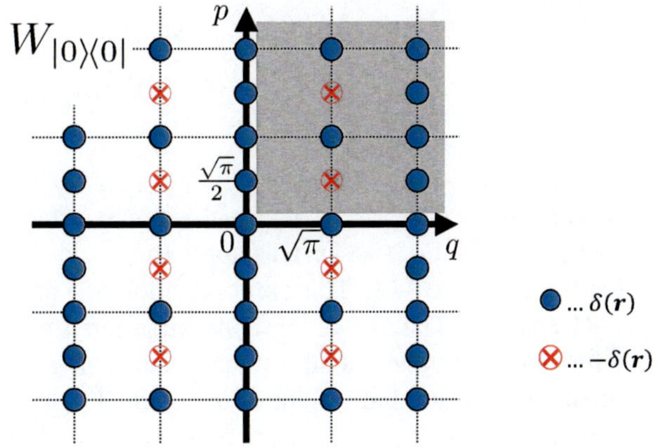

Fig. 5.1 The Wigner function of the ideal GKP state $|0^{(\text{ideal})}\rangle$, where each blue filled circle represents a positive delta function $\delta(\boldsymbol{r})$ with $\boldsymbol{r} = (q, p)$ and each red circled X represents a negative delta function $-\delta(\boldsymbol{r})$. This Wigner function has periodicity, where the gray region shows a period

$$W_{|j^{(\text{ideal})}\rangle\langle j^{(\text{ideal})}|}(q, p)$$

$$= \frac{1}{2}\sum_{\substack{t\in\mathbb{Z}\\ t'\in\mathbb{Z}}} e^{-\pi i t t'}\delta\left(p - \frac{\alpha_d t}{2}\right)\delta\left(q - \frac{\alpha_d d t'}{2} - \alpha_d j\right) \tag{5.32}$$

$$= \frac{1}{2}\sum_{\substack{t\in\mathbb{Z}\\ t'\in\mathbb{Z}}} \delta\left(p + \frac{\alpha_d t}{2}\right)\left[\delta\left(q - \alpha_d d\left(t' + \frac{j}{d}\right)\right) + (-1)^t\,\delta\left(q - \alpha_d d\left(t' + \frac{j}{d} + \frac{1}{2}\right)\right)\right]. \tag{5.33}$$

Figure 5.1[2] shows the schematics of the ideal GKP logical basis state $|0^{(\text{ideal})}\rangle$. This shows that the Wigner function of the ideal logical basis states forms a square lattice consisting of Dirac delta functions, which has half the period of the Dirac comb in the position and momentum representations. Since its sublattice formed of the odd periods starting from $(q, p) = (\alpha_d j, 0)$ consists of the Dirac delta functions with negative signs, the comb for the odd periods in position cancels out when integrated over momentum, and vice versa.

5.3 On the Equivalence of Approximate Gottesman-Kitaev-Preskill Codes

As defined so far, the ideal GKP-encoded states are non-normalizable and thus unphysical. (All the relations described so far are formal.) The ideal GKP code should be regarded as a limit of physically meaningful approximate codes. Various approximations of the GKP-encoded states are considered in the past literature [1, 17, 31, 42, 50–53]. The following three approximations are conventionally used.

[2] Modified from [45], Copyright 2020 by American Physical Society.

(Approximation 1)

$$|j_{\kappa,\Delta}^{(1)}\rangle := \frac{1}{\sqrt{N_{\kappa,\Delta,j}^{(1)}}} \sum_{s\in\mathbb{Z}} e^{-\frac{1}{2}\kappa^2\alpha_d^2(ds+j)^2} X(\alpha_d(ds+j)) S(-\ln\Delta)\,|0\rangle_f, \tag{5.34}$$

where $\kappa, \Delta > 0$, $N_{\kappa,\Delta,j}^{(1)}$ is a normalization constant, and $S(\xi)$ is the squeezing operator defined in Eq. (3.84). This approximate code state approaches the ideal one as the limit of $\kappa, \Delta \to 0$. This approximation first appeared in the original paper of the GKP code [1]. The idea of this approximation is to replace the position "eigenstates" with squeezed coherent states with a squeezing parameter $\ln(1/\Delta)$ and superpose them with a Gaussian weight of the width $1/\kappa$. This gives us an insight on how to generate the GKP-encoded state experimentally [54].

(Approximation 2)

$$|j_{\gamma,\delta}^{(2)}\rangle := \frac{1}{\sqrt{N_{\gamma,\delta,j}^{(2)}}} \iint \frac{dr_1 dr_2}{2\pi\gamma\delta}\, e^{-\frac{r_1^2}{2\gamma^2}-\frac{r_2^2}{2\delta^2}}\, V(\boldsymbol{r})|j^{(\mathrm{ideal})}\rangle, \tag{5.35}$$

where γ and δ satisfy $0 < \gamma\delta < 2$, and $V(\boldsymbol{r}) := \exp(-ir_1r_2/2)Z(r_1)X(r_2)$. This approximate code state approaches the ideal one as $\gamma, \delta \to 0$. This approximation also appeared in the original paper to regard the approximation as an error, and treat $\frac{1}{2\pi\gamma\delta}e^{-\frac{r_1^2}{2\gamma^2}-\frac{r_2^2}{2\delta^2}}$ as an error "wave function" [1]. They use the term "wave function" because the state given in Eq. (5.35) is not an ideal code state subject to the error caused by the random displacement channel, but a coherent superposition of randomly displaced ideal code states. The error "wave function" later turned out to have more profound meanings; it is actually a wave function in the "grid representation" [18, 53, 55–57], which is an analogous representation to the position representation, but with respect to the so-called "shifted grid states" instead of position eigenstates. In Appendix A.1, we make remarks on the "grid representation" in terms of the representation theory of the Heisenberg group.

(Approximation 3)

$$|j_{\beta}^{(3)}\rangle := \frac{1}{\sqrt{N_{\beta,j}^{(3)}}} e^{-\beta(\hat{n}+\frac{1}{2})}|j^{(\mathrm{ideal})}\rangle, \tag{5.36}$$

where β satisfies $\beta > 0$, and it approaches the ideal code state as $\beta \to 0$. Contrary to the former two approximations, Approximation 3, first appearing in Ref. [17], only deals with symmetric envelope in position and momentum coordinates. Since the approximation factor $e^{-\beta(\hat{n}+\frac{1}{2})}$ is diagonal in the Fock basis, this approximation may be useful for computing the statistical properties of operators which are diagonal in the Fock basis, as shown in Ref. [17]. On the other hand, though this approximate code state could conceptually be prepared by feeding the ideal code states to the

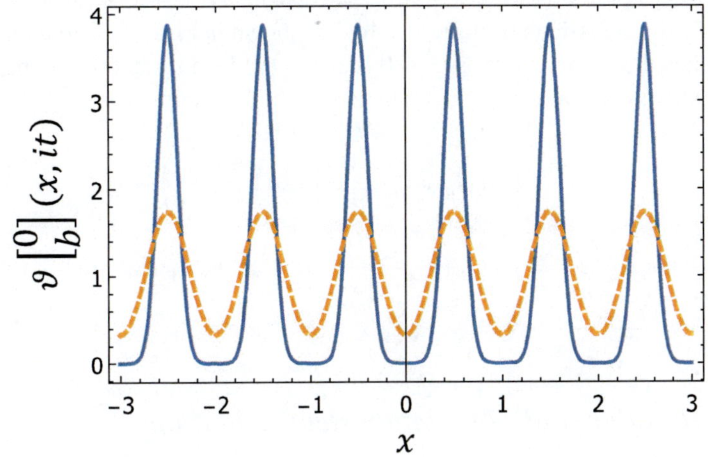

Fig. 5.2 The theta function in the form of Eq. (5.39) with respect to x when $b = 1/2$ and $t = 1/15$ (blue solid line), and when $b = 1/2$ and $t = 1/3$ (yellow dashed line). The theta function in this form is a sequence of the same Gaussian functions with respect to x which has peaks at $b, b \pm 1, b \pm 2, \ldots$, and the width of each Gaussian is determined by t as shown in the figure. Note that Eq. (5.39) approaches the Dirac comb as $t \to 0$

beamsplitter followed by post-selecting the vacuum click at the idler port [42], it provides few implications about their realistic experimental generation.

In this section, we prove that all the three approximations listed above are equivalent up to the squeezing. Before going into the proofs, we summarize the relevant functions used in this chapter. For $z \in \mathbb{C}$ and $\tau \in \mathbb{C}$ satisfying $\mathrm{Im}(\tau) > 0$, let $\vartheta(z, \tau) := \sum_{s\in\mathbb{Z}} \exp(\pi i \tau s^2 + 2\pi i z s)$ be the theta function (we follow the notation in Ref. [58]), and

$$\vartheta\begin{bmatrix} a \\ b \end{bmatrix}(z, \tau) := \sum_{s\in\mathbb{Z}} \exp[\pi i \tau (s+a)^2 + 2\pi i (z+b)(s+a)] \tag{5.37}$$

$$= \exp[\pi i \tau a^2 + 2\pi i a(z+b)]\, \vartheta(z + \tau a + b, \tau) \tag{5.38}$$

be the theta function with rational characteristics (a, b) [58]. The theta functions which we mainly use are in the form

$$\vartheta\begin{bmatrix} 0 \\ b \end{bmatrix}(x, it), \tag{5.39}$$

where $x, t \in \mathbb{R}$, and $b \in \mathbb{Q}$. The theta function in this form is a sequence of the same Gaussian functions with respect to x which has peaks at $b, b \pm 1, b \pm 2, \ldots$, and the width of each Gaussian is determined by t as shown in Fig. 5.2.[3] Note that Eq. (5.39) approaches the Dirac comb as $t \to 0$. Let $G_{\sigma^2}(x)$ be a probability density function of the normal distribution with variance σ^2, which is defined as

$$G_{\sigma^2}(x) := \frac{1}{\sqrt{2\pi\sigma^2}} \exp\left(-\frac{x^2}{2\sigma^2}\right). \tag{5.40}$$

Finally, let $f * g(x) := \int dy\, f(y)g(x-y)$ denote the convolution of two functions $f(x)$ and $g(x)$.

5.3.1 *Position and Momentum Representations*

In order to determine the relationship among the three approximations, we derive the position and momentum representations of the approximate code states. Note that the position and momentum representations of Approximation 1 have already appeared in the past literature [1, 18, 32, 50, 52–54, 56, 59, 60], but we rewrite them for completeness. For this purpose, we define the following functions.

Definition 5.3.1 Define $E_{\mu,\Gamma,a}(x)$ and $\tilde{E}_{\mu,\Gamma,a}(x)$ as

$$E_{\mu,\Gamma,a}(x) := \sum_{s\in\mathbb{Z}} \exp\left(-\frac{(s+a)^2\Gamma^2}{2\mu}\right) \delta(x-(s+a)\Gamma)\,, \tag{5.41}$$

$$\tilde{E}_{\mu,\Gamma,a}(x) := \sum_{s\in\mathbb{Z}} \exp\left(-\frac{s^2\Gamma^2}{2\mu} + 2\pi i a s\right) \delta(x+s\Gamma)\,. \tag{5.42}$$

The function $E_{\mu,\Gamma,a}(x)$ is a Dirac comb with its interval given by Γ, which is shifted by the rational a of the interval from the origin and weighted by the Gaussian $\exp(-x^2/(2\mu))$ of the width μ. It can also be interpreted as a Fourier transform of the theta function in the form of $\frac{1}{\sqrt{2\pi}}\vartheta\begin{bmatrix} a \\ 0 \end{bmatrix}\left(\frac{\Gamma}{2\pi}x, \frac{i\Gamma^2}{2\pi\mu}\right)$ with respect to x, which can be confirmed by its definition Eq. (5.37). On the other hand, the function $\tilde{E}_{\mu,\Gamma,a}(x)$, a Dirac comb with the Gaussian weight which has a phase factor for each peak, is a Fourier transform of the theta function in the form of $\frac{1}{\sqrt{2\pi}}\vartheta\begin{bmatrix} 0 \\ a \end{bmatrix}\left(-\frac{\Gamma}{2\pi}x, \frac{i\Gamma^2}{2\pi\mu}\right)$, which can also be confirmed by Eq. (5.37).

Now, under Definition 5.3.1, we show the following proposition.

Proposition 5.3.2 (The position representation) *Let* $\kappa, \Delta, \beta > 0$ *and* $0 < \gamma\delta < 2$. *Define* $\lambda(\gamma,\delta) := 1 + \frac{\gamma^2\delta^2}{4}$. *Then, the position representations of the states Eqs.* (5.34), (5.35), *and* (5.36) *are given as follows:*

[3] Modified from [44], Copyright 2020 by American Physical Society.

- *(Approximation 1)*

$$\langle q|j_{\kappa,\Delta}^{(1)}\rangle = \left(\frac{2\sqrt{\pi}\Delta^2}{N_{\kappa,\Delta,j}^{(1)}}\right)^{\frac{1}{2}} E_{\frac{1}{\kappa^2},\alpha_d d,\frac{j}{d}} * G_{\Delta^2}(q) \tag{5.43}$$

$$= \left(\frac{2\sqrt{\pi}\Delta^2}{\kappa^2 d N_{\kappa,\Delta,j}^{(1)}}\right)^{\frac{1}{2}} G_{\frac{1+\kappa^2\Delta^2}{\kappa^2}}(q)\, \vartheta\begin{bmatrix} 0 \\ j/d \end{bmatrix}\left(-\frac{q}{\alpha_d d(1+\kappa^2\Delta^2)}, \frac{i\Delta^2}{d(1+\kappa^2\Delta^2)}\right). \tag{5.44}$$

- *(Approximation 2)*

$$\langle q|j_{\gamma,\delta}^{(2)}\rangle = \left(\frac{\alpha_d d}{\lambda(\gamma,\delta) N_{\gamma,\delta,j}^{(2)}}\right)^{\frac{1}{2}} E_{\frac{\lambda(\gamma,\delta)}{\gamma^2}\left(1-\frac{\gamma^2\delta^2}{2\lambda(\gamma,\delta)}\right)^2,\,\alpha_d d\left(1-\frac{\gamma^2\delta^2}{2\lambda(\gamma,\delta)}\right),\,\frac{j}{d}} * G_{\frac{\delta^2}{\lambda(\gamma,\delta)}}(q) \tag{5.45}$$

$$= \left(\frac{\alpha_d \gamma^{-2}}{N_{\gamma,\delta,j}^{(2)}}\right)^{\frac{1}{2}} G_{\frac{\lambda(\gamma,\delta)}{\gamma^2}}(q)\, \vartheta\begin{bmatrix} 0 \\ j/d \end{bmatrix}\left(-\frac{q}{\alpha_d d}\left[1-\frac{\gamma^2\delta^2}{2\lambda(\gamma,\delta)}\right], \frac{i\delta^2}{d\lambda(\gamma,\delta)}\right). \tag{5.46}$$

- *(Approximation 3)*

$$\langle q|j_{\beta}^{(3)}\rangle = \left(\frac{\alpha_d d}{\cosh\beta\, N_{\beta,j}^{(3)}}\right)^{\frac{1}{2}} E_{\frac{1}{\sinh\beta\cosh\beta},\,\frac{\alpha_d d}{\cosh\beta},\,\frac{j}{d}} * G_{\tanh\beta}(q) \tag{5.47}$$

$$= \left(\frac{\alpha_d}{\sinh\beta\, N_{\beta,j}^{(3)}}\right)^{\frac{1}{2}} G_{\frac{1}{\tanh\beta}}(q)\, \vartheta\begin{bmatrix} 0 \\ j/d \end{bmatrix}\left(-\frac{q}{\alpha_d d\cosh\beta}, \frac{i\tanh\beta}{d}\right). \tag{5.48}$$

For each approximation, we gave the two expressions in which we smear the Dirac delta functions in the definition of the ideal GKP-encoded state with the Gaussian functions in different orders. In the expressions (5.43), (5.45), and (5.47), each peak of the Dirac comb, which is weighted by a Gaussian as shown in the definition of $E_{\mu,\Gamma,a}$, is convoluted with another Gaussian $G_\nu(q)$. In the alternative expressions (5.44), (5.46), and (5.48), the infinite sequence of Gaussian spikes as defined in $\vartheta\begin{bmatrix} 0 \\ a \end{bmatrix}(q, it)$ is multiplied by another Gaussian function $G_{\nu'}(q)$ which works as an overall envelope. The expressions (5.43), (5.45), and (5.47) are suited for understanding the physical structure of the approximation such as the interval of the neighboring Gaussian peaks. The alternative expressions (5.44), (5.46), and (5.48) are convenient for numerical calculations because algorithms to calculate the theta function with arbitrary precision are well known [61].

Sketch of the proof. We derive Eqs. (5.43), (5.45), and (5.47) with straightforward but cumbersome calculations, and then apply the following lemma to derive Eqs. (5.44), (5.46), and (5.48).

Lemma 5.3.3 *For $\mu, \nu > 0$, $\Gamma \in \mathbb{R}$, and $a \in \mathbb{Q}$, the following equality holds:*

$$E_{\mu,\Gamma,a} * G_\nu(q) = \sqrt{\frac{2\pi\mu}{\Gamma^2}}\, G_{\mu+\nu}(q)\, \vartheta\begin{bmatrix}0\\a\end{bmatrix}\left(-\frac{q}{(1+\nu/\mu)\Gamma}, \frac{2\pi i\nu}{(1+\nu/\mu)\Gamma^2}\right), \tag{5.49}$$

$$\tilde{E}_{\mu,\Gamma,a} * G_\nu(q) = \sqrt{\frac{2\pi\mu}{\Gamma^2}}\, G_{\mu+\nu}(q)\, \vartheta\begin{bmatrix}a\\0\end{bmatrix}\left(-\frac{q}{(1+\nu/\mu)\Gamma}, \frac{2\pi i\nu}{(1+\nu/\mu)\Gamma^2}\right) \tag{5.50}$$

The full proof of Proposition 5.3.2 as well as the proof of Lemma 5.3.3 is in Appendix A.2.1. □

Under Definition 5.3.1, the momentum representations of the approximate code states can also be given by the following corollary.

Corollary 5.3.4 (The momentum represenation) *Let $\kappa, \Delta, \beta > 0$ and $0 < \gamma\delta < 2$. Let $\lambda(\gamma,\delta) := 1 + \frac{\gamma^2\delta^2}{4}$. Then, the momentum representations of the states (5.34), (5.35), and (5.36) are given as follows:*

- *(Approximation 1)*

$$\langle p|j^{(1)}_{\kappa,\Delta}\rangle = \left(\frac{2\sqrt{\pi\Delta^2}}{(1+\kappa^2\Delta^2)dN^{(1)}_{\kappa,\Delta,j}}\right)^{\frac{1}{2}} \tilde{E}_{\frac{1}{\Delta^2(1+\kappa^2\Delta^2)}, \frac{\alpha_d}{1+\kappa^2\Delta^2}, \frac{j}{d}} * G_{\frac{\kappa^2}{1+\kappa^2\Delta^2}}(p). \tag{5.51}$$

- *(Approximation 2)*

$$\langle p|j^{(2)}_{\gamma,\delta}\rangle = \left(\frac{\alpha_d}{\lambda(\gamma,\delta)N^{(2)}_{\gamma,\delta,j}}\right)^{\frac{1}{2}} \tilde{E}_{\frac{\lambda(\gamma,\delta)}{\delta^2}\left(1-\frac{\gamma^2\delta^2}{2\lambda(\gamma,\delta)}\right)^2, \alpha_d\left(1-\frac{\gamma^2\delta^2}{2\lambda(\gamma,\delta)}\right), \frac{j}{d}} * G_{\frac{\gamma^2}{\lambda(\gamma,\delta)}}(p). \tag{5.52}$$

- *(Approximation 3)*

$$\langle p|j^{(3)}_{\beta}\rangle = \left(\frac{\alpha_d}{\cosh\beta\, N^{(3)}_{\beta,j}}\right)^{\frac{1}{2}} \tilde{E}_{\frac{1}{\sinh\beta\cosh\beta}, \frac{\alpha_d}{\cosh\beta}, \frac{j}{d}} * G_{\tanh\beta}(p). \tag{5.53}$$

Here, we only write the expressions in terms of $\tilde{E}_{\mu,\Gamma,a}$, but the expressions in terms of the theta function can also be obtained by applying Lemma 5.3.3 to these expressions.

Proof We use the fact that the momentum representation of a state is a Fourier transform of its position representation, i.e., $\langle p|j\rangle = \frac{1}{\sqrt{2\pi}}\int dq\, e^{-ipq}\langle q|j\rangle$. We can thus derive Eqs. (5.51), (5.52), and (5.53) as Fourier transforms of Eqs. (5.44), (5.46), and (5.48), respectively, exploiting the fact that the Fourier transform of the product of two functions is given by the convolution of the Fourier transforms of the respective functions, and the Fourier transform of $\frac{1}{\sqrt{2\pi}}\vartheta\begin{bmatrix}0\\a\end{bmatrix}\left(-\frac{\Gamma}{2\pi}x, \frac{i\Gamma^2}{2\pi\mu}\right)$ is $\tilde{E}_{\mu,\Gamma,a}$ while the Fourier transform of G_ν is $\sqrt{1/\nu}\, G_{\frac{1}{\nu}}$.

5.3.2 *Explicit Relations Among the Three Approximations*

The position and momentum representations of the three different approximate GKP logical basis states lead to conditions for equivalence of these approximations. Since $E_{\mu,\Gamma,a} * G_\nu(x)$ denotes the array of the Gaussian spikes $G_\nu(x)$ at intervals Γ, one can notice from Eqs. (5.45) and (5.47) that the intervals of the Gaussian spikes of the approximate code states are narrower than those of the ideal one, $\alpha_d d$, in the case of Approximations 2 and 3. Furthermore, from Eqs. (5.52) and (5.53), the intervals of the Gaussian spikes of each of these approximate code states in the momentum representations get narrower in the same proportion as that of their respective position representations. With this observation, Approximation 3, which has symmetric envelope functions in position and momentum representations Eqs. (5.47) and (5.53), is expected to be a symmetric case ($\gamma = \delta$) of Approximation 2 in the sense of "symmetric" in Definition 5.2.1. This can be confirmed by the following.

Corollary 5.3.5 (The symmetric code) *Let $R(\pi/2)$ be the Fourier operator defined in Eq. 3.83. Then, the following relation holds for the logical basis states of the Approximation 3:*

$$R(-\pi/2)|j_\beta^{(3)}\rangle = \sum_{j'=0}^{d-1}\sqrt{\frac{N_{\beta,j'}^{(3)}}{N_{\beta,j}^{(3)}}}|j_\beta'^{(3)}\rangle. \tag{5.54}$$

The same relation holds for Approximation 2 iff $\gamma = \delta$, i.e.,

$$R(-\pi/2)|j_{\gamma,\gamma}^{(2)}\rangle = \sum_{j'=0}^{d-1}\sqrt{\frac{N_{\gamma,\gamma,j'}^{(2)}}{N_{\gamma,\gamma,j}^{(2)}}}|j_{\gamma,\gamma}'^{(2)}\rangle. \tag{5.55}$$

Proof It can be observed by combining $\langle x|_{\hat{q}} R(-\pi/2) = \langle x|_{\hat{p}}$ with Eqs. (5.45), (5.47), (5.46), and (5.53).

In contrast with Approximations 2 and 3, the interval of Gaussian spikes in the position representation (5.43) of Approximation 1 is the same as that in the position representation of the ideal code state, and the interval in the momentum representation (5.51) of Approximation 1 is narrower than that in the momentum representation of the ideal code state. This means that Approximation 1 narrows the lattice spacing of the code space asymmetrically in position and momentum. This suggests that Approximation 1 may be related to Approximation 2 and Approximation 3 by a transformation that symmetrizes the interval of the lattice spacing in position and momentum.

We confirm this by applying the squeezing operation $S(\ln\sqrt{1+\kappa^2\Delta^2})$ for symmetrizing the intervals of the Gaussian spikes of the code state $|j^{(1)}_{\kappa,\Delta}\rangle$ in position and momentum coordinates:

$$\langle q|_{\hat{q}} S\left(\ln\sqrt{1+\kappa^2\Delta^2}\right)|j^{(1)}_{\kappa,\Delta}\rangle = (1+\kappa^2\Delta^2)^{\frac{1}{4}}\langle\sqrt{1+\kappa^2\Delta^2}q|j^{(1)}_{\kappa,\Delta}\rangle \tag{5.56}$$

$$= \sqrt{m}\, E_{\frac{1}{\kappa^2(1+\kappa^2\Delta^2)}, \frac{\alpha_d d}{\sqrt{1+\kappa^2\Delta^2}}, \frac{j}{d}} * G_{\frac{\Delta^2}{(1+\kappa^2\Delta^2)}}(q), \tag{5.57}$$

$$\langle p|_{\hat{p}} S\left(\ln\sqrt{1+\kappa^2\Delta^2}\right)|j^{(1)}_{\kappa,\Delta}\rangle = (1+\kappa^2\Delta^2)^{-\frac{1}{4}}\langle p/\sqrt{1+\kappa^2\Delta^2}|j^{(1)}_{\kappa,\Delta}\rangle \tag{5.58}$$

$$= \sqrt{\frac{m}{d}}\, \tilde{E}_{\frac{1}{\Delta^2}, \frac{\alpha_d}{\sqrt{1+\kappa^2\Delta^2}}, \frac{j}{d}} * G_{\kappa^2}(p), \tag{5.59}$$

where $m = \frac{2}{N^{(1)}_{\kappa,\Delta,j}}\sqrt{\frac{\pi\Delta^2}{1+\kappa^2\Delta^2}}$. In order to derive Eqs. (5.57) and (5.59), we used $E_{\mu,\Gamma,a} * G_\nu(bx) = \frac{1}{b}E_{\frac{\mu}{b^2},\frac{\Gamma}{b},a} * G_{\frac{\nu}{b^2}}(x)$ and $\tilde{E}_{\mu,\Gamma,a} * G_\nu(bx) = \frac{1}{b}\tilde{E}_{\frac{\mu}{b^2},\frac{\Gamma}{b},a} * G_{\frac{\nu}{b^2}}(x)$, which can be obtained from the definition of the functions $E_{\mu,\Gamma,a}(x)$, $\tilde{E}_{\mu,\Gamma,a}(x)$, and $G_\nu(x)$. Comparing the position representation (5.57) of the squeezed version of Approximation 1 with the position representation (5.45) of Approximation 2 and (5.47) of Approximation 3, we arrive at the following theorem.

Theorem 5.3.6 (Equivalence of the approximate GKP logical basis states) *By choosing the parameters in Approximations 1 and 2 as*

$$\kappa^2 = \frac{\gamma^2}{\lambda(\gamma,\delta)} = \tanh\beta, \tag{5.60}$$

$$\Delta^2 = \frac{\delta^2}{\lambda(\gamma,\delta)}\left(1-\frac{\gamma^2\delta^2}{2\lambda(\gamma,\delta)}\right)^{-2} = \sinh\beta\cosh\beta, \tag{5.61}$$

$$\gamma^2 = \delta^2 = 2\tanh\frac{\beta}{2}, \tag{5.62}$$

where $\lambda(\gamma,\delta) := 1+\frac{\gamma^2\delta^2}{4}$, *we have*

$$S\left(\ln\sqrt{1+\kappa^2\Delta^2}\right)|j^{(1)}_{\kappa,\Delta}\rangle = |j^{(2)}_{\gamma,\delta}\rangle = |j^{(3)}_{\beta}\rangle. \tag{5.63}$$

Proof It directly follows from Eqs. (5.45), (5.47), and (5.57). □

Theorem 5.3.6 together with Corollary 5.3.5 shows that up to a squeezing of the factor $\ln(\sqrt{1+\kappa^2\Delta^2})$ for Approximation 1 in order to make the code symmetric in the sense of Definition 5.2.1, the logical basis states of the symmetric code of Approximations 1, 2, and 3 are exactly the same quantum state. This squeezing becomes negligible in the limit of good approximation. In this sense, all these approximations are equivalent up to a squeezing that is ignorable in the limit of good approximation. This definition of equivalence is well-motivated since single-mode Gaussian unitary operations are easy to implement compared to non-Gaussian operations on continuous-variable systems such as optical systems, and among displacement, phase rotation, and squeezing for decomposing Gaussian unitaries [3], only squeezing can change the lattice spacing.

The converse of the theorem is also true; the choice of parameters in Theorem 5.3.6 is the only choice for the logical basis states of these approximations to be the same quantum states. This fact can be seen by the following remark.

Remark 5.3.7 So far, we followed the convention to fix the lattice spacing parameter as $\alpha = \alpha_d$, and derived equivalence relations among symmetric approximate codes. Such an exact correspondence between approximate codes can be generalized to asymmetric case. Let us remove the constraint of Eq. (5.14) and regard α as a free parameter in each approximation, and define states $|j^{(1)}_{\kappa,\Delta,\alpha}\rangle$, $|j^{(2)}_{\gamma,\delta,\alpha}\rangle$, and $|j^{(3)}_{\beta,\alpha}\rangle$ (see Appendix A.2.1. We can observe from Eqs. (A.13) and (A.20) in Appendix A.2.1 that $|j^{(1)}_{\kappa,\Delta,\alpha}\rangle = |j^{(2)}_{\gamma,\delta,\alpha'}\rangle$ with the following choice of parameters:

$$\kappa^2 = \frac{\gamma^2}{\lambda(\gamma,\delta)}\left(1 - \frac{\gamma^2\delta^2}{2\lambda(\gamma,\delta)}\right)^{-2}, \tag{5.64}$$

$$\alpha = \alpha'\left(1 - \frac{\gamma^2\delta^2}{2\lambda(\gamma,\delta)}\right), \tag{5.65}$$

$$\Delta^2 = \frac{\delta^2}{\lambda(\gamma,\delta)}. \tag{5.66}$$

Compared to $|j^{(1)}_{\kappa,\Delta,\alpha}\rangle$ and $|j^{(2)}_{\gamma,\delta,\alpha}\rangle$, the third approximation $|j^{(3)}_{\beta,\alpha}\rangle$ has fewer parameters and cannot always be made equivalent to $|j^{(1)}_{\kappa,\Delta,\alpha}\rangle$ and $|j^{(2)}_{\gamma,\delta,\alpha}\rangle$. It is because each Gaussian spike of the third approximation $|j^{(3)}_{\beta,\alpha}\rangle$ always has the same variance in position and momentum. If we apply the squeezing $S(\ln\zeta)$ to $|j^{(3)}_{\beta,\alpha}\rangle$ so that the variances of Gaussian spike in position and momentum can differ, we have $|j^{(1)}_{\kappa,\Delta,\alpha}\rangle = |j^{(2)}_{\gamma,\delta,\alpha'}\rangle = S(\ln\zeta)|j^{(3)}_{\beta,\alpha''}\rangle$ with the following correspondence of the parameters in addition to Eqs. (5.64), (5.65), and (5.66):

$$\kappa^2 = \zeta^2 \sinh\beta\cosh\beta, \tag{5.67}$$

$$\alpha = \frac{\alpha''}{\zeta\cosh\beta}, \tag{5.68}$$

$$\Delta^2 = \frac{\tanh\beta}{\zeta^2}. \tag{5.69}$$

This can be confirmed from the fact that $\langle q|S(\ln\zeta)|j\rangle = \sqrt{\zeta}\langle\zeta q|j\rangle$, and $E_{\mu,\Gamma,a} * G_\nu(\zeta q) = \zeta^{-1} E_{\frac{\mu}{\zeta^2},\frac{\Gamma}{\zeta},a} * G_{\frac{\nu}{\zeta^2}}(q)$.

Remark 5.3.8 The equivalence of Approximation 2 with $\gamma = \delta$ and Approximation 3 can also be proved from Eqs. (1.4) and (7.12) in Ref. [31] by setting $l = l' = 0$, while Ref. [31] does not prove the equivalence. Our contribution here is to derive their position wave functions in Proposition 5.3.2 and to show the equivalence using these position wave functions.

5.3.3 The Standard Form

Now that we have shown the equivalence of Approximations 1, 2, and 3, we introduce a standard form of the approximate GKP logical basis state, which we will use in the rest of the paper.

Definition 5.3.9 (*Standard form of the approximate GKP logical basis states*) Given three parameters σ_q^2, σ_p^2, and Γ with $0 < \sigma_q^2, \sigma_p^2 < 1/2$, the standard form of the approximate GKP code is defined as the code which is spanned by a logical qudit basis $\{|j_{\sigma_q^2,\sigma_p^2,\Gamma}\rangle : j = 0, \ldots, d-1\}$ with its position representation given by

$$\langle q|j_{\sigma_q^2,\sigma_p^2,\Gamma}\rangle := \left(\frac{2\Gamma\left(\Lambda(\sigma_q^2,\sigma_p^2)\right)^{-\frac{1}{2}}}{N_{\sigma_q^2,\sigma_p^2,\Gamma,j}}\right)^{\frac{1}{2}} E_{\frac{\Lambda(\sigma_q^2,\sigma_p^2)}{2\sigma_p^2},\Gamma,\frac{j}{d}} * G_{2\sigma_q^2}(q), \tag{5.70}$$

where $\Lambda(\sigma_q^2, \sigma_p^2)$ is defined as

$$\Lambda(\sigma_q^2,\sigma_p^2) := 1 - 4\sigma_q^2\sigma_p^2, \tag{5.71}$$

and $N_{\sigma_q^2,\sigma_p^2,\Gamma,j}$ is a normalization constant. For the symmetric code, the logical basis $\{|j_{\sigma^2}\rangle : j = 0, \ldots, d-1\}$ is parametrized by only one parameter σ^2 $(0 < \sigma^2 < 1/2)$ as

$$\langle q|j_{\sigma^2}\rangle := \left(\frac{2\alpha_d d}{N_{\sigma^2,j}}\right)^{\frac{1}{2}} E_{\frac{\Lambda(\sigma^2)}{2\sigma^2},\alpha_d d\sqrt{\Lambda(\sigma^2)},\frac{j}{d}} * G_{2\sigma^2}(q), \tag{5.72}$$

where

$$\Lambda(\sigma^2) := 1 - 4\sigma^4. \tag{5.73}$$

Note that $|j_{\sigma^2}\rangle$ is equal to $|j_{\sigma^2,\sigma^2,\alpha_d d\sqrt{\Lambda(\sigma^2)}}\rangle$. The momentum representation of $|j_{\sigma_q^2,\sigma_p^2,\Gamma}\rangle$ is given by

$$\langle p|j_{\sigma_q^2,\sigma_p^2,\Gamma}\rangle = \left(\frac{4\pi\sqrt{\Lambda(\sigma_q^2,\sigma_p^2)}}{\Gamma N_{\sigma_q^2,\sigma_p^2,\Gamma,j}}\right)^{\frac{1}{2}} \tilde{E}_{\frac{\Lambda(\sigma_q^2,\sigma_p^2)}{2\sigma_q^2},\frac{2\pi\Lambda(\sigma_q^2,\sigma_p^2)}{\Gamma},\frac{j}{d}} * G_{2\sigma_p^2}(p), \tag{5.74}$$

and thus, for the symmetric code, it is given by

$$\langle p|j_{\sigma^2}\rangle = \left(\frac{2\alpha_d}{N_{\sigma^2,j}}\right)^{\frac{1}{2}} \tilde{E}_{\frac{\Lambda(\sigma^2)}{2\sigma^2},\alpha_d\sqrt{\Lambda(\sigma^2)},\frac{j}{d}} * G_{2\sigma^2}(p). \tag{5.75}$$

We can also write Eqs. (5.70), (5.72), (5.74), and (5.75) in terms of the theta function by using Lemma 5.3.3.

The physical meanings of the parameters σ_q^2, σ_p^2, and Γ of the state $|j_{\sigma_q^2,\sigma_p^2,\Gamma}\rangle$ (and hence σ^2 of the state $|j_{\sigma^2}\rangle$) will be clarified in Sect. 5.3.4. Furthermore, an explicit form of the normalization constant $N_{\sigma_q^2,\sigma_p^2,\Gamma,j}$ (and hence $N_{\sigma^2,j}$) is given in Proposition 5.3.11. The expressions corresponding to Approximations 1–3 can be obtained simply by substituting the corresponding parameters:

$$|j^{(1)}_{\kappa,\Delta}\rangle = |j_{\sigma_q^2=\frac{\Delta^2}{2},\sigma_p^2=\frac{\kappa^2}{2(1+\Delta^2)},\Gamma=\alpha_d d}\rangle, \tag{5.76}$$

$$|j^{(2)}_{\gamma,\delta}\rangle = |j_{\sigma_q^2=\frac{\delta^2}{2\lambda(\gamma,\delta)},\sigma_p^2=\frac{\gamma^2}{2\lambda(\gamma,\delta)},\Gamma=\alpha_d d\left(1-\frac{\gamma^2\delta^2}{2\lambda(\gamma,\delta)}\right)}\rangle, \tag{5.77}$$

$$|j^{(3)}_{\beta}\rangle = |j_{\sigma^2=\frac{\tanh\beta}{2}}\rangle, \tag{5.78}$$

where $\lambda(\gamma,\delta) = 1 + \frac{\gamma^2\delta^2}{4}$.

5.3.4 *Explicit Expressions of the Wigner Function, Inner Products, and Average Photon Number*

In this section, we derive the expressions of the Wigner function, inner products, and the average photon number for the standard form of the approximate code state $|j_{\sigma_q^2,\sigma_p^2,\Gamma}\rangle$ in Definition 5.3.9. Those for $|j_{\sigma^2}\rangle$ can also be obtained by substituting $\sigma_q^2 = \sigma_p^2 = \sigma^2$ and $\Gamma = \alpha_d d\sqrt{\Lambda(\sigma^2)}$. They also have expressions in terms of the Riemann theta function [58] (also known as the Siegel theta function), which is a multi-variable generalization of the theta function. These alternative expressions are relatively neat, but for the later analyses of the asymptotic behaviors (5.89) and (5.90), the expressions in terms of the theta function are more convenient. Thus we give the alternative expressions in terms of the Riemann theta function in Appendix A.3.

We will first derive the Wigner function of the operators $|j_{\sigma_q^2,\sigma_p^2,\Gamma}\rangle\langle j'_{\sigma_q^2,\sigma_p^2,\Gamma}|$. The Wigner function of the approximate GKP code can be used for the analyses of quantum error correction, as shown in Refs. [17, 33, 34].

Proposition 5.3.10 (Wigner function) *For the approximate code states* $|j_{\sigma_q^2,\sigma_p^2,\Gamma}\rangle$ *and* $|j'_{\sigma_q^2,\sigma_p^2,\Gamma}\rangle$ *in Definition 5.3.9, the Wigner function* $W_{|j_{\sigma_q^2,\sigma_p^2,\Gamma}\rangle\langle j'_{\sigma_q^2,\sigma_p^2,\Gamma}|}(q,p)$ *of the operator* $|j_{\sigma_q^2,\sigma_p^2,\Gamma}\rangle\langle j'_{\sigma_q^2,\sigma_p^2,\Gamma}|$ *is given by*

$$
\begin{aligned}
&W_{|j_{\sigma_q^2,\sigma_p^2,\Gamma}\rangle\langle j'_{\sigma_q^2,\sigma_p^2,\Gamma}|}(q,p)\\
&= \frac{1}{\sqrt{N_{\sigma_q^2,\sigma_p^2,\Gamma,j} N_{\sigma_q^2,\sigma_p^2,\Gamma,j'}}} \left[\left(E_{\frac{\Lambda(\sigma_q^2,\sigma_p^2)}{4\sigma_p^2},\Gamma,\frac{j+j'}{2d}} * G_{\sigma_q^2}(q) \right) \left(\tilde{E}_{\frac{\Lambda(\sigma_q^2,\sigma_p^2)}{4\sigma_q^2},\frac{\pi\Lambda(\sigma_q^2,\sigma_p^2)}{\Gamma},\frac{j-j'}{2d}} * G_{\sigma_p^2}(p) \right) \right.\\
&\qquad \left. + \left(E_{\frac{\Lambda(\sigma_q^2,\sigma_p^2)}{4\sigma_p^2},\Gamma,\frac{j+j'}{2d}+\frac{1}{2}} * G_{\sigma_q^2}(q) \right) \left(\tilde{E}_{\frac{\Lambda(\sigma_q^2,\sigma_p^2)}{4\sigma_q^2},\frac{\pi\Lambda(\sigma_q^2,\sigma_p^2)}{\Gamma},\frac{j-j'}{2d}+\frac{1}{2}} * G_{\sigma_p^2}(p) \right) \right],
\end{aligned} \tag{5.79}
$$

where E and $\tilde{E}$ are defined in Definition 5.3.1, and $\Lambda(\sigma_q^2, \sigma_p^2)$ is defined in Eq. (5.71).

Calculations for deriving the Wigner function are similar to those for deriving the position and momentum representations but are more complicated. The proof of Proposition 5.3.10 is in Appendix A.2.2. We can also write Eq. (5.79) in terms of the theta function by applying Lemma 5.3.3 to Eq. (A.43) in Appendix A.2.2.

The Wigner function in Proposition 5.3.10 shows the physical meanings of σ_q^2, σ_p^2, and Γ. The first term in the square bracket of Eq. (5.79) with $j = j'$ denotes an infinite sequence of Gaussian spikes each of which has variance σ_q^2 in position and σ_p^2 in momentum with periods Γ and $\pi\Lambda(\sigma_q^2, \sigma_p^2)\Gamma^{-1}$, respectively, and has overall Gaussian envelopes with variances $(4\sigma_p^2)^{-1}\Lambda(\sigma_q^2, \sigma_p^2)$ and $(4\sigma_q^2)^{-1}\Lambda(\sigma_q^2, \sigma_p^2)$, respectively. The second term shows that the same structure is also at the places shifted by half periods in position, but with positive and negative signs alternately in momentum. The Gaussian spikes in the first and second terms with different signs interfere destructively when projected onto position or momentum, while constructively with the same signs. Since $E_{\mu,\Gamma,a}(x) \to \sum_{s\in\mathbb{Z}} \delta(x - (s+a)\Gamma)$ and $\tilde{E}_{\mu,\Gamma,a}(x) \to \sum_{s\in\mathbb{Z}} e^{2\pi i a s} \delta(x + s\Gamma)$ as $\mu \to \infty$, and $G_\nu(x) \to \delta(x)$ as $\nu \to 0$, we can observe that Eq. (5.79) with $\Gamma = \alpha_d d$ approaches Eq. (5.33) as $\sigma_q^2, \sigma_p^2 \to 0$, as expected. Note that for the symmetric code state $|j_{\sigma^2}\rangle$, the intervals of the neighboring Gaussian peaks $\alpha_d d\sqrt{\Lambda(\sigma^2)}$ is smaller than $\alpha_d d$ of the ideal one. The change in the interval is $\mathcal{O}(\sigma^4)$, and thus may be negligible for small σ^2. However, in experiments, we cannot always make σ^2 small enough to keep the change in the intervals negligible. With our results, we can quantitatively analyze the code performance even for not necessarily small σ^2.

Using Eq. (5.79) with the explicit form of the normalization constant given in Proposition 5.3.11, we plot the Wigner function of the GKP logical basis state in Figure 5.3.[4] Note that a similar expression has already been used in Ref. [17] with a more intuitive explanation. Our contribution here is to derive the Wigner function corresponding to the approximate code states explicitly, which we will use in the detailed analysis of the average photon number.

Next, using the Wigner function (5.79), we provide a closed-form expression for the normalization constant $N_{\sigma_q^2,\sigma_p^2,\Gamma}$. The normalization constants were calculated

[4] Modified from [44], Copyright 2020 by American Physical Society.

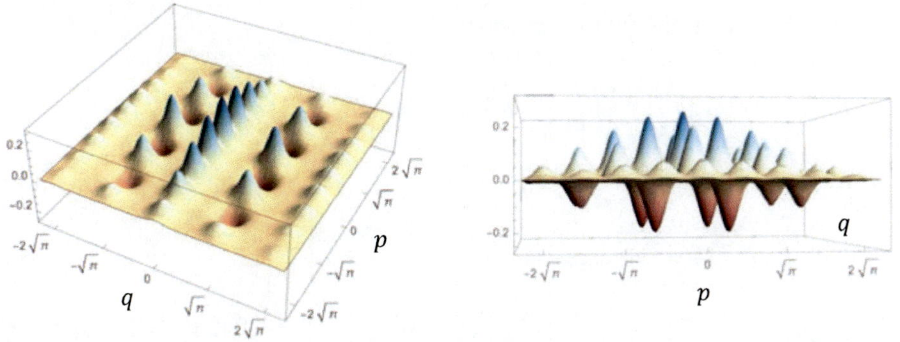

Fig. 5.3 The Wigner function of the symmetric code state $|0_{\sigma^2}\rangle$ in the case $d = 2$ with $\sigma^2 = 0.05$, which is calculated from Eq. (5.79) along with the explicit form of the normalization constant given in Proposition 5.3.11

numerically in previous works [17, 31, 56, 62], but here we can provide their analytical expressions in terms of the theta functions. Furthermore, since the logical basis states of the approximate GKP codes are nonorthogonal, their inner products are nonzero in general, which we quantitatively analyze in the following. Since the theta functions used in the following proposition can be calculated with arbitrary precision by a method in, e.g., Ref. [61], the results are useful for evaluating the code performance reliably, as demonstrated in Sect. 5.4.

Proposition 5.3.11 (Normalization constant and inner product) *The normalization factor* $N_{\sigma_q^2,\sigma_p^2,\Gamma,j}$ *of the approximate code state* $|j_{\sigma_q^2,\sigma_p^2,\Gamma}\rangle$ *in Definition 5.3.9 is given in terms of the theta functions by*

$$N_{\sigma_q^2,\sigma_p^2,\Gamma,j} = \vartheta\begin{bmatrix}\frac{j}{d}\\0\end{bmatrix}\left(0, \frac{2i\Gamma^2\sigma_p^2}{\pi\Lambda(\sigma_q^2,\sigma_p^2)}\right)\vartheta\begin{bmatrix}0\\0\end{bmatrix}\left(0, \frac{2\pi i\sigma_q^2\Lambda(\sigma_q^2,\sigma_p^2)}{\Gamma^2}\right) + \vartheta\begin{bmatrix}\frac{j}{d}+\frac{1}{2}\\0\end{bmatrix}\left(0, \frac{2i\Gamma^2\sigma_p^2}{\pi\Lambda(\sigma_q^2,\sigma_p^2)}\right)\vartheta\begin{bmatrix}0\\\frac{1}{2}\end{bmatrix}\left(0, \frac{2\pi i\sigma_q^2\Lambda(\sigma_q^2,\sigma_p^2)}{\Gamma^2}\right). \tag{5.80}$$

Furthermore, the inner product between $|j_{\sigma_q^2,\sigma_p^2,\Gamma}\rangle$ *and another approximate code state* $|j'_{\sigma_q^2,\sigma_p^2,\Gamma}\rangle$ *is given by*

$$\begin{aligned}&\langle j'_{\sigma_q^2,\sigma_p^2,\Gamma}|j_{\sigma_q^2,\sigma_p^2,\Gamma}\rangle\\ &= \frac{1}{\sqrt{N_{\sigma_q^2,\sigma_p^2,\Gamma,j}N_{\sigma_q^2,\sigma_p^2,\Gamma,j'}}}\left\{\vartheta\begin{bmatrix}\frac{j+j'}{2d}\\0\end{bmatrix}\left(0, \frac{2i\Gamma^2\sigma_p^2}{\pi\Lambda(\sigma_q^2,\sigma_p^2)}\right)\vartheta\begin{bmatrix}0\\\frac{j-j'}{2d}\end{bmatrix}\left(0, \frac{2\pi i\sigma_q^2\Lambda(\sigma_q^2,\sigma_p^2)}{\Gamma^2}\right)\right.\\ &\left.+\vartheta\begin{bmatrix}\frac{j+j'}{2d}+\frac{1}{2}\\0\end{bmatrix}\left(0, \frac{2i\Gamma^2\sigma_p^2}{\pi\Lambda(\sigma_q^2,\sigma_p^2)}\right)\vartheta\begin{bmatrix}0\\\frac{j-j'}{2d}+\frac{1}{2}\end{bmatrix}\left(0, \frac{2\pi i\sigma_q^2\Lambda(\sigma_q^2,\sigma_p^2)}{\Gamma^2}\right)\right\}.\end{aligned} \tag{5.81}$$

Proof We exploit the following facts:

$$\langle j'|j\rangle = \mathrm{Tr}\left[|j\rangle\langle j'|\right] = \iint dqdp\, W_{|j\rangle\langle j'|}(q, p), \tag{5.82}$$

$$\int dx\, f * g(x) = \int dx\, f(x) \int dy\, g(y), \tag{5.83}$$

$$\int dx\, E_{\mu,\Gamma,a}(x) = \sum_{s\in\mathbb{Z}} \exp[-(s+a)^2\Gamma^2/2\mu] = \vartheta\begin{bmatrix} a \\ 0 \end{bmatrix}\left(0, \frac{i\Gamma^2}{2\pi\mu}\right), \tag{5.84}$$

$$\int dx\, \tilde{E}_{\mu',\Gamma',a'}(x) = \sum_{s\in\mathbb{Z}} \exp[-\Gamma'^2 s^2/2\mu' + 2\pi i a' s] = \vartheta\begin{bmatrix} 0 \\ a' \end{bmatrix}\left(0, \frac{i\Gamma'^2}{2\pi\mu'}\right), \tag{5.85}$$

$$\int dx\, G_{\sigma^2}(x) = 1. \tag{5.86}$$

Combining the above with the Wigner function of $W_{|j_{\sigma_q^2,\sigma_p^2,\Gamma}\rangle\langle j'_{\sigma_q^2,\sigma_p^2,\Gamma}|}$ in Eq. (5.79), we obtain Eqs. (5.80) and (5.81).

The expressions in Proposition 5.3.11 are exact and applicable to any σ^2, but at the same time complicated. Thus, we investigate their asymptotic behaviors in order to obtain intuitive relations with respect to the degree of approximation. As shown in Ref. [63], the asymptotic behavior of the theta function in the form of $\vartheta\begin{bmatrix} a \\ 0 \end{bmatrix}(0, it)$ as $t \to +0$ is given by

$$\vartheta\begin{bmatrix} a \\ 0 \end{bmatrix}(0, it) = \sum_{s=0}^{\infty} e^{-\pi t(s+a)^2} + \sum_{s=0}^{\infty} e^{-\pi t(s+1-a)^2} = \frac{1}{\sqrt{t}} + \mathcal{O}\left(t^{\frac{1}{2}}\right). \tag{5.87}$$

Furthermore, the asymptotic behavior of $\vartheta\begin{bmatrix} 0 \\ a \end{bmatrix}(0, it)$ as $t \to +0$ is given by

$$\vartheta\begin{bmatrix} 0 \\ a \end{bmatrix}(0, it) = \frac{1}{\sqrt{t}}\vartheta\begin{bmatrix} a \\ 0 \end{bmatrix}(0, it^{-1}) = \frac{1}{\sqrt{t}} \sum_{s=-\infty}^{\infty} e^{-\frac{\pi}{t}(s+a)^2} \simeq \frac{1}{\sqrt{t}} e^{-\frac{\pi}{t}a^2}, \tag{5.88}$$

where we used Eq. (A.34) in Appendix A.2.1 in the first equality. The last approximation takes the dominant term as $t \to +0$ up to a constant coefficient.

Now the asymptotic form of the logarithm of the normalization constant $N_{\sigma_q^2,\sigma_p^2,\Gamma,j}$ in Eq. (5.80) as $\sigma_q^2, \sigma_p^2 \to +0$ is given by

$$\ln N_{\sigma_q^2,\sigma_p^2,\Gamma,j} \to -\ln\sqrt{4\sigma_q^2\sigma_p^2}. \tag{5.89}$$

In the same way, the asymptotic behavior of the logarithm of $|\langle j'_{\sigma_q^2,\sigma_p^2,\Gamma}|j_{\sigma_q^2,\sigma_p^2,\Gamma}\rangle|$ in Eq. (5.81) as $\sigma_q^2, \sigma_p^2 \to +0$ is given by

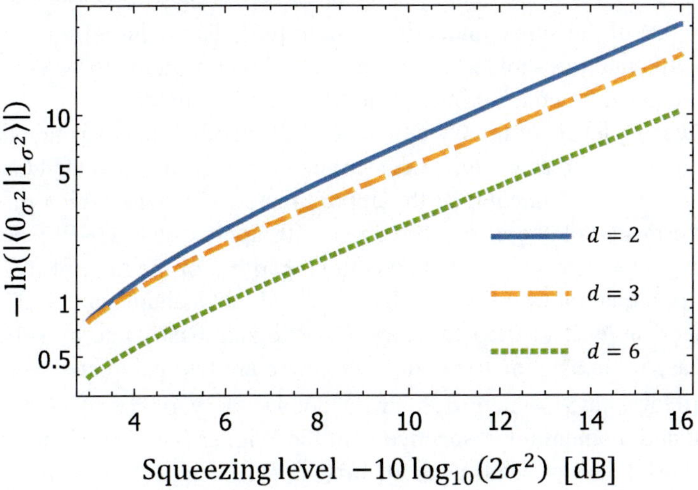

Fig. 5.4 The logarithms of the absolute values of an inner product $-\ln\left(\left|\langle 0_{\sigma^2}|1_{\sigma^2}\rangle\right|\right)$ for the code state (5.72) in Definition 5.3.9 with $d = 2,\ 3$, and 6. The horizontal axis, $-10\log_{10}(2\sigma^2)$, is a squeezing level in decibels, which is a convention to express the degree of squeezing. The vertical axis is in the log scale. One can observe that, in the region where the squeezing level is over 5 dB for $d = 3$ and 6, the negative logarithm of the inner product increases linearly with respect to the squeezing level in the log plot, that is, $-\ln\left|\langle 0_{\sigma^2}|1_{\sigma^2}\rangle\right| \propto \sigma^{-2}$, as expected in the asymptotic behavior (5.90)

$$\ln|\langle j'_{\sigma_q^2,\sigma_p^2,\Gamma}|j_{\sigma_q^2,\sigma_p^2,\Gamma}\rangle| \to -\frac{(j'-j)^2\Gamma^2}{8d^2\sigma_q^2}. \tag{5.90}$$

The overlap between logical basis states thus decreases exponentially with respect to σ_q^{-2}.

Along with the asymptotic behavior, we numerically calculate Eqs. (5.80) and (5.81) to see how the overlaps between code states change with respect to the degree of approximation. Figure 5.4[5] shows the logarithms of the absolute values of an inner product $\left|\langle 0_{\sigma^2}|1_{\sigma^2}\rangle\right|$ of the approximate code states (5.72) in Definition 5.3.9 with $d = 2, 3$, and 6, with respect to a squeezing level in decibels $-10\log_{10}(2\sigma^2)$, which is a quality measure of an approximate code state and explained later. One can observe that, in the region where the squeezing level is over 5 dB for $d = 3$ and 6, the minus of the logarithm of the inner product increases linearly with respect to the squeezing level in the log plot, that is, $-\ln\left|\langle 0_{\sigma^2}|1_{\sigma^2}\rangle\right| \propto \sigma^{-2}$, as expected in the asymptotic behavior (5.90). In the case of $d = 2$, the inclination of the plot is larger than those in the case of $d = 3$ and 6, which may be caused by a constant factor in Eq. (5.90) when $\frac{j+j'}{2d} \lesssim \frac{1}{2}$. Note that the squeezing levels of the GKP-encoded states when $d = 2$ in the recent experiments are 5.5–7.3 dB with the position and momentum degrees of freedom in trapped ion system [64], and 7.4–9.5 dB with the

[5] Modified from [44], Copyright 2020 by American Physical Society.

cavity mode of the superconducting system [65]. The required squeezing level for the fault-tolerant threshold of the universal QC is considered to be 8–16 dB [34–36, 38, 40], depending on experimental setups and noise models.

"Squeezing level" of the (symmetric) GKP-encoded state was first considered in Ref. [17] in order to characterize the variance σ^2 of each convoluted Gaussian spike G_{σ^2} in the Wigner function of the approximate code state, which directly affects the performance of the error correction with approximate GKP codes. Since the squeezing level of a squeezed state is the logarithm of the ratio of the variances of the position quadrature $(\Delta\hat{q})^2$ of that state and the vacuum state, Ref. [17] defines the squeezing level of the symmetric GKP logical basis state by $-10\log_{10}(2\sigma^2)$. In the case of an asymmetric code state, there are two parameters $-10\log_{10}(2\sigma_q^2)$ and $-10\log_{10}(2\sigma_p^2)$, where σ_q^2 and σ_p^2 denote the variance of Gaussian spike in position and momentum, respectively, in the Wigner function of the standard form (5.79). This definition leads us to identifying $-10\log_{10}\Delta^2$ ($\simeq -10\log_{10}\kappa^2$) as the squeezing parameter for Approximation 1 [33–36, 65] due to the relation (5.76). Note that there also exists another definition of "effective squeezing parameter", motivated by quantum metrology [53, 57, 64]. In this paper, we adopt the former definition as a "squeezing level" in order to observe the relation between the performance of error correction and the average photon number of the approximate code states.

Finally, using the Wigner function (5.79) of the approximate code state $|j_{\sigma_q^2,\sigma_p^2,\Gamma}\rangle$, we can calculate the average photon number of the code state. Below we write $\langle A\rangle_{|j_{\sigma_q^2,\sigma_p^2,\Gamma}\rangle} := \langle j_{\sigma_q^2,\sigma_p^2,\Gamma}|A|j_{\sigma_q^2,\sigma_p^2,\Gamma}\rangle$ for an operator A.

Proposition 5.3.12 (Average photon number) *The average photon number $\langle\hat{n}\rangle_{|j_{\sigma_q^2,\sigma_p^2,\Gamma}\rangle}$ of the approximate code state $|j_{\sigma_q^2,\sigma_p^2,\Gamma}\rangle$ in Definition 5.3.9 is given as follows:*

$$\langle\hat{n}\rangle_{|j_{\sigma_q^2,\sigma_p^2,\Gamma}\rangle} = \frac{\sigma_q^2+\sigma_p^2-1}{2} - \left(\frac{\partial}{\partial x}+\frac{\partial}{\partial y}\right)\ln\tilde{N}_{\sigma_q^2,\sigma_p^2,\Gamma,j}(x,y)\Big|_{x=\frac{4\sigma_p^2}{\Lambda(\sigma_q^2,\sigma_p^2)},\ y=\frac{4\sigma_q^2}{\Lambda(\sigma_q^2,\sigma_p^2)}}, \tag{5.91}$$

where $\tilde{N}_{\sigma_q^2,\sigma_p^2,\Gamma,j}(x,y)$ is defined as

$$\begin{aligned}\tilde{N}_{\sigma_q^2,\sigma_p^2,\Gamma,j}(x,y) := {} & \vartheta\begin{bmatrix}\frac{j}{d}\\0\end{bmatrix}\left(0,\frac{i\Gamma^2}{2\pi}x\right)\vartheta\begin{bmatrix}0\\0\end{bmatrix}\left(0,\frac{\pi i[\Lambda(\sigma_q^2,\sigma_p^2)]^2}{2\Gamma^2}y\right)\\ & +\vartheta\begin{bmatrix}\frac{j}{d}+\frac{1}{2}\\0\end{bmatrix}\left(0,\frac{i\Gamma^2}{2\pi}x\right)\vartheta\begin{bmatrix}0\\\frac{1}{2}\end{bmatrix}\left(0,\frac{\pi i[\Lambda(\sigma_q^2,\sigma_p^2)]^2}{2\Gamma^2}y\right)\end{aligned} \tag{5.92}$$

Sketch of proof. Using the Wigner function (5.79), we can derive the expectation values of the square of the position and momentum quadrature, $\langle\hat{q}^2\rangle_{|j_{\sigma_q^2,\sigma_p^2,\Gamma}\rangle}$ and $\langle\hat{p}^2\rangle_{|j_{\sigma_q^2,\sigma_p^2,\Gamma}\rangle}$. Then we can derive the explicit expression of $\langle\hat{n}\rangle_{|j_{\sigma_q^2,\sigma_p^2,\Gamma}\rangle}$ by exploiting the fact that $\langle \hat{q}^2 + \hat{p}^2 \rangle_{|j_{\sigma_q^2,\sigma_p^2,\Gamma}\rangle} = \langle 2\hat{n} + 1 \rangle_{|j_{\sigma_q^2,\sigma_p^2,\Gamma}\rangle}$. The full proof is in Appendix A.2.3. □

As an application of the results, we observe the relation between the squeezing level and the average photon number of approximate code states. The squeezing level has a direct connection to the performance of the quantum error correction using GKP codes [17, 33–36, 40]. On the other hand, the average photon number of the encoded state is relevant to the capacity of the continuous-variable quantum channel [4, 42, 66], which works as an effective dimension of the Hilbert space. Since it is found that the GKP code has high performance in the channel coding for bosonic Gaussian channels [31, 42], the connections between these two notions are important for further analyses of the Gaussian channel coding.

Previous literature estimates the average photon number of the encoded state as $\simeq \frac{1}{4\sigma^2} - \frac{1}{2}$ for the symmetric code for given squeezing level $-10\log_{10}(2\sigma^2) \gg 1$ [1, 17, 31, 42, 51, 56]. This is because the variance of the envelope Gaussian in the Wigner function of the approximate code states is roughly equal to $\frac{1}{4\sigma^2}$, and the average photon number relates to the expectation values of the squares of the position and momentum quadratures by $\langle \hat{q}^2 + \hat{p}^2 \rangle_{|j_{\sigma^2}\rangle} = \langle 2\hat{n} + 1 \rangle_{|j_{\sigma^2}\rangle}$. It is also consistent with the expression of the average photon number given in Eq. (5.91) when the asymptotic form $\tilde{N}_{\sigma_q^2,\sigma_p^2,\Gamma,j}(x, y) \propto \frac{1}{\sqrt{xy}}$ is considered. However, this estimation is no longer valid in the case of a low squeezing level. Here we are interested in the squeezing level at which this estimation deviates from the exact value.

We compute the average photon number of the code state $|j_{\sigma^2}\rangle$ defined in Eq. (5.72) in Definition 5.3.9 with $d = 2$, by using the formula (5.91). As mentioned above, the squeezing level of $|j_{\sigma^2}\rangle$ is given by $-10\log_{10}(2\sigma^2)$. Fig. 5.5[6] shows the average photon number of $|0_{\sigma^2}\rangle$ and $|1_{\sigma^2}\rangle$ with respect to the squeezing level $-10\log_{10}(2\sigma^2)$. In Fig. 5.5, we compare our result with a conventionally used estimate of the average photon number $\frac{1}{4\sigma^2} - \frac{1}{2}$. The figure reveals that when the squeezing level is less than 10 dB, the conventionally used estimate of the average photon number deviates from the exact values. Note that 10 dB squeezing is in a range of the expected thresholds for fault-tolerant continuous-variable QC [34–36, 38, 40], which is a curious coincidence.

5.3.5 Discussion

In this section, we explicitly showed conditions under which the conventional approximations of the GKP code, Approximations 1, 2, and 3, defined in Eqs. (5.34), (5.35), and (5.36), are made equivalent. We observed that up to a slight squeezing for Approximation 1, Approximations 1, 2, and 3 are equivalent for the symmetric code, in which the logical basis states and their Fourier transforms span the same code space. Furthermore, we quantitatively showed that in all these approximations, the lattice spacing of the Gaussian spikes in phase space appearing in the description

[6] Modified from [44], Copyright 2020 by American Physical Society.

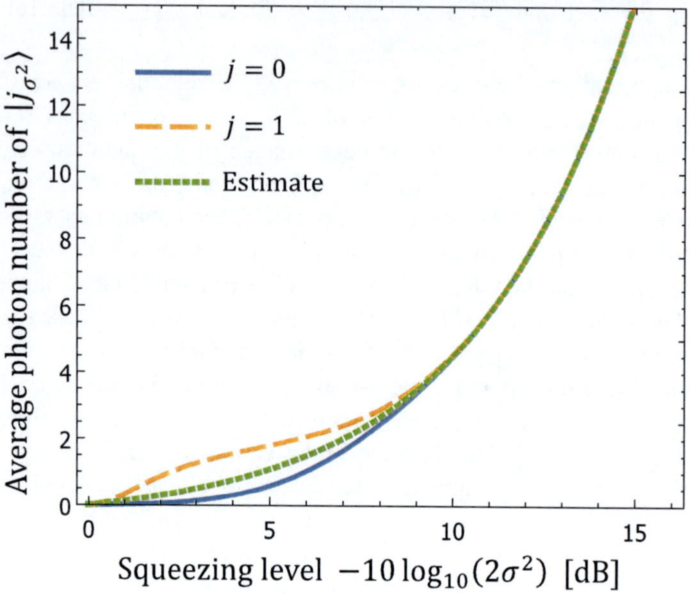

Fig. 5.5 The average photon number of the code state (5.72) in Definition 5.3.9 with $d = 2$. "Estimate" denotes the function $\frac{1}{4\sigma^2} - \frac{1}{2}$. These three are in good accordance when the squeezing level is over 10 dB, but our rigorous calculations provide better estimates at the squeezing levels in the recent experiments, that is, 5.5–7.3 dB in the trapped ion system [64] and 7.4–9.5 dB in the superconducting system [65]

of the approximate code states is narrower than that of the corresponding ideal GKP code state. Although this effect may be negligible in the limit of large squeezing levels, it potentially affects the performance of the error correction. It is because the error correction strategy explicitly depends on the lattice spacing of the code states in the GKP code. Quantitatively, in the case of approximate code state of $d = 2$ with 8 dB squeezing, the lattice spacing is about 1% narrower than that of the ideal one. It is thus needed to investigate error correction schemes taking the change in lattice spacing into account, especially at a moderate squeezing level relevant to experimental realizations of GKP codes.

Exploiting the equivalence, we also gave the standard form of the approximate code states in terms of the position representation. Furthermore, we derived the explicit formulas of the Wigner function, normalization constant, inner product, and the average photon number of the logical basis states. These tools given in this section may accelerate further theoretical developments of continuous-variable quantum information processing based on quantum error correction and channel coding with the GKP error-correcting code.

5.4 Cost-Reduced All-Gaussian Universality with the GKP Code

In the QC with qubits (or more generally qudits), the Clifford operations, i.e., the operations realized by the combinations of the Pauli-eigenstate preparations, the Clifford gates, and the Pauli measurements, cannot realize the universal QC [48, 49]. This no-go statement is known as the Gottesman-Knill theorem. In order to perform the universal QC, one needs to implement a non-Clifford gate [2] such as the T-gate defined as

$$T = \begin{pmatrix} e^{-\frac{\pi}{8}i} & 0 \\ 0 & e^{\frac{\pi}{8}i} \end{pmatrix}. \tag{5.93}$$

The same is true for the GKP code. For the GKP code, the GKP-logic Clifford gates, i.e., the logical Clifford gates on qubits encoded in the continuous-variable system by the GKP code, can be realized only by Gaussian operations (defined in Sect. 3.1.4) as stated in the previous section. In order to realize the universality, a GKP-logic non-Clifford gate is necessary. In the original paper [1], the GKP-logic non-Clifford gate is shown to be implementable either by using the optical non-Gaussian operation or by using the GKP-encoded magic states such as $|H\rangle$ and $|\frac{\pi}{8}\rangle$, where these magic states are defined as

$$|H\rangle := \cos(\pi/8)|0\rangle + \sin(\pi/8)|1\rangle, \tag{5.94}$$

$$|\tfrac{\pi}{8}\rangle := (e^{-\frac{\pi}{8}i}|0\rangle + e^{\frac{\pi}{8}i}|1\rangle)/\sqrt{2}. \tag{5.95}$$

Here we remark that Gaussian operations and logical Clifford operations on the GKP qubits are different in that Pauli eigenstates of the GKP code, such as $|0\rangle$ and $|1\rangle$, are non-Gaussian; i.e., initialization of GKP qubits requires non-Gaussian operations.

Reference [43] has recently shown that the noisy GKP-encoded magic state can be probabilistically generated if one has the ability to prepare the GKP $|0\rangle$ states and to perform Gaussian operations. It is known that a higher-fidelity magic state can be distilled from many low-fidelity ones if the fidelity of lower ones is above a certain threshold. Therefore, the proposed method of probabilistically preparing noisy $|H\rangle$ means that we need only one type of a GKP-encoded state $|0\rangle$ with Gaussian operations to realize the universal QC. It is no longer required to develop technologies for preparing both $|0\rangle$ and $|H\rangle$ of the GKP code. However, in contrast to the majority of qubit-based codes in which the preparation of the magic state is much more costly than the preparation of the Pauli eigenstate, both $|0\rangle$ and $|H\rangle$ of the GKP code are non-Gaussian, and hence the preparation of them is equally costly compared to the realization of Gaussian operations in the quantum optical system. Thus, the overhead of consuming many expensive $|0\rangle$s for the distillation of $|H\rangle$ may become a bottleneck for the optical quantum computer.

In order to overcome the obstacle arising from the magic state distillation and achieve a fundamental cost reduction in implementing optical QC, we propose an idea of preparing only the GKP-logic magic state $|H\rangle$. We show a scheme that realizes the universal QC by combining Gaussian operations only with $|H\rangle$ instead of $|0\rangle$. Compared to the previous scheme, our scheme does not suffer from the overhead of the magic state distillation because $|0\rangle$ can be deterministically prepared from only two $|H\rangle$s by the state-injection protocol [67]. While the scheme itself is pretty simple, our contribution is a fundamental cost reduction of non-Gaussian resources for implementing the optical quantum computer. Furthermore, we introduce a simple method for obtaining a fundamental limit on the transformation between $|H\rangle$ and $|0\rangle$ of the GKP code by any Gaussian operations, utilizing the resource theory of non-Gaussianity [68, 69], one of quantum resource theories [70] for continuous-variable QC. We also show the feasibility of the direct preparation of $|H\rangle$ of the GKP code. The existing proposals [39, 50, 52–54, 56, 59, 62, 64, 65, 71–85] for the preparation of the GKP-encoded state mostly focus on the preparation of $|0\rangle$ state, but we discuss possibilities to generalize these proposals to the preparation of $|H\rangle$ state with a comparable technological cost.

5.4.1 Deterministic All-Gaussian Universality Using the GKP Magic States

In the following, the logical qubit encoded in a physical continuous-variable system by the GKP code is referred to as *a GKP qubit*, and a physical state of a GKP qubit as *a GKP state*. Towards realizing fault-tolerant QC with quantum optical systems, it is promising to combine Gaussian operations with the GKP qubits [17, 34, 35, 38]. It is because Gaussian errors on the continuous-variable systems, which frequently occur in quantum optical systems, cannot be corrected solely by Gaussian operations [86] but can be by combining Gaussian operations with an approximate GKP code concatenated with a multi-qubit quantum error-correcting code [17].

The GKP code can be used not only for correcting errors on the continuous-variable system but also for realizing the universal QC with its non-Gaussianity. The protocol developed in Ref. [43] that realizes the universal QC utilizing the non-Gaussianity of the GKP-encoded state is based on the magic state distillation [67, 87] that probabilistically and approximately transforms GKP qubits prepared in $|0\rangle \otimes |0\rangle \otimes \cdots$ into a magic state $|H\rangle$ only using Gaussian operations. This protocol suggests that when Gaussian operations are available, the ability to prepare *only one type* of the GKP-encoded state $|0\rangle$ suffices to implement universal QC. However, this protocol has the significant overhead arising from the magic state distillation; whenever we need to obtain one GKP magic state $|H\rangle$ with a sufficiently high fidelity, many $|0\rangle$s will be consumed. More precisely, the required number of $|0\rangle$s to prepare a GKP magic state $|H\rangle$ up to an infidelity ϵ with the scheme in Ref. [43] amounts to [88–90]

$$\mathcal{O}\left(\text{polylog}\left(\frac{1}{\epsilon}\right)\right) \quad \text{as } \epsilon \to 0. \tag{5.96}$$

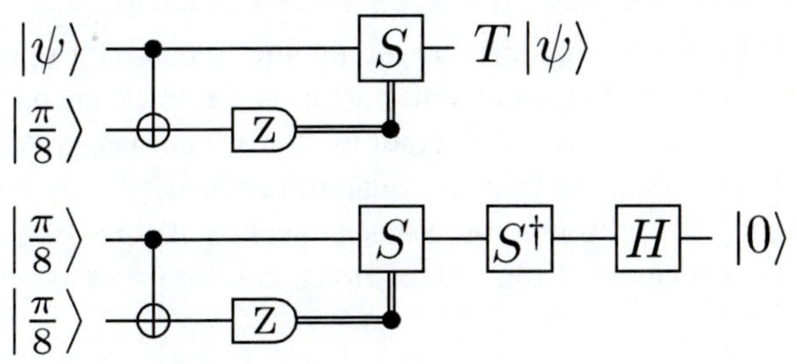

Fig. 5.6 A quantum circuit of state injection for applying the T gate to any one-qubit input state $|\psi\rangle$ by Clifford operations assisted by an auxiliary input qubit prepared in $|\frac{\pi}{8}\rangle$ at the top, and that for converting a two-qubit input state $|\frac{\pi}{8}\rangle^{\otimes 2}$ to $|0\rangle$ at the bottom. The latter conversion circuit can be implemented only by adaptive Gaussian operations on the GKP qubits, namely, performing the Clifford gates (CNOT, S, $S^\dagger$, and H) that are implemented with Gaussian unitary operations, and conditioning on an outcome of the Z-basis measurement that is implemented with a homodyne detection

This overhead per the preparation of $|H\rangle$ pushes up the implementation cost of the fault-tolerant logical non-Clifford gate on a qubit-based quantum error-correcting code concatenated with the GKP code and thus raise the total cost of the fault-tolerant QC.

In order to reduce the mass consumption of GKP qubits, here we propose choosing $|H\rangle$ instead of $|0\rangle$ as the single GKP state along with Gaussian operations for realizing the universal QC. Note that $|H\rangle$ and $|\frac{\pi}{8}\rangle$ defined in Eqs. (5.94) and (5.95) are Clifford equivalent, i.e., related by $|H\rangle = SH|\frac{\pi}{8}\rangle$ with the Clifford gates H and S given in Eqs. (5.2) and (5.3), which can thus be interchanged by Gaussian operations in the GKP code. Therefore, in the following we use these states interchangeably. Our cost-reduced scheme use a well-known quantum circuit for the T-gate teleportation [1, 2, 67] given at the top of Fig. 5.6.[7] This circuit can apply the T gate to an arbitrary input state $|\psi\rangle$ by the combination of an ancillary magic state $|\frac{\pi}{8}\rangle$ and Clifford operations. When we set $|\psi\rangle = |\frac{\pi}{8}\rangle$, we have

$$T|\tfrac{\pi}{8}\rangle = e^{-\pi i/4}(|0\rangle + i|1\rangle)/\sqrt{2} = e^{-\pi i/4} SH|0\rangle, \tag{5.97}$$

and can thus prepare the GKP $|0\rangle$ with additional Gaussian operations as shown at the bottom of Figure 5.6. We can thus deterministically transform two GKP qubits prepared in $|\frac{\pi}{8}\rangle^{\otimes 2}$ into $|0\rangle$ only by Gaussian operations. Our deterministic protocol transforming two $|H\rangle$ states to a $|0\rangle$ state can be advantageous over the probabilistic protocol developed in Ref. [43] transforming many $|0\rangle$ states to a $|H\rangle$ state; given the target infidelity ϵ, the number of the consumed GKP-encoded states in our protocol is bounded by a constant in contrast to (5.96), which essentially reduces the overhead. In fact, in the ideal case, $|0\rangle$ state can be exactly ($\epsilon = 0$) prepared in our protocol while it cannot be in Ref. [43].

[7] Modified from [45], Copyright 2020 by American Physical Society.

Notice that this cost reduction may be unique to quantum optical architectures in which preparing non-Gaussian states such as the GKP-encoded states is more costly than performing Gaussian operations. It does not necessarily hold for other continuous-variable systems than the quantum optical system, such as superconducting cavities [65] and trapped-ion mechanical oscillators [64], where Gaussian operations are not necessarily reliable to implement compared to non-Gaussian operations.

5.4.2 A Resource-Theoretical Analyses for Fundamental Limitations on GKP State Conversion

Since the convertibility between the GKP states $|H\rangle$ and $|0\rangle$ under Gaussian operations is crucial for the argument so far, we here develop a simple method for obtaining fundamental bounds on the convertibility between the GKP states. For this analysis, we use the resource theory of non-Gaussianity in which Gaussian operations are considered to be free and non-Gaussian states, unitaries, and measurements are regarded as resources for assisting Gaussian operations [68, 69, 91, 92]. Following the rule of the convex resource theories [68, 69], we include adaptive Gaussian operations conditioned on the outcomes of Gaussian measurements in the free operations. Note that the resource theories of magic [93, 94], where Clifford operations are considered to be free, are inapplicable to our case because Pauli eigenstates of the GKP qubits, e.g., $|0\rangle$, are the resourceful states while they are free in the resource theories of magic.

Resource theories of non-Gaussianity introduce a measure that quantifies the non-Gaussianity of a state on the continuous-variable system. We may be able to convert a state with high non-Gaussianity into that with low non-Gaussianity but cannot in the reverse direction. One way to quantify the non-Gaussianity of a given state ψ is to use the logarithmic negativity [68, 69] of the Wigner function W_ψ of ψ defined as $\mathcal{N}_L(\psi) := \ln\left(\iint_{-\infty}^{\infty} dqdp \, \left|W_\psi(q, p)\right|\right)$. For the pure Gaussian state ρ, for example, we have $\mathcal{N}_L(\rho) = 0$ since the Wigner function W_ρ of the pure Gaussian state ρ is always nonnegative [95, 96]. The logarithmic negativity $\mathcal{N}_L$ does not increase under any Gaussian operations (monotonicity).

Although the logarithmic negativity is well-defined for the Wigner functions of approximate GKP states, it is not for the Wigner functions of the ideal GKP states, where infinitely many Dirac delta functions are arranged according to a square lattice in concept, as depicted in Fig. 5.7.[8] We can compute the logarithmic negativity for the Wigner functions of approximate GKP states in principle, but we would like more intuitive ways of comparing the non-Gaussianity and analyzing the convertibility between ideal GKP states. For this, we develop an alternative measure to quantify

[8] Modified from [45], Copyright 2020 by American Physical Society.

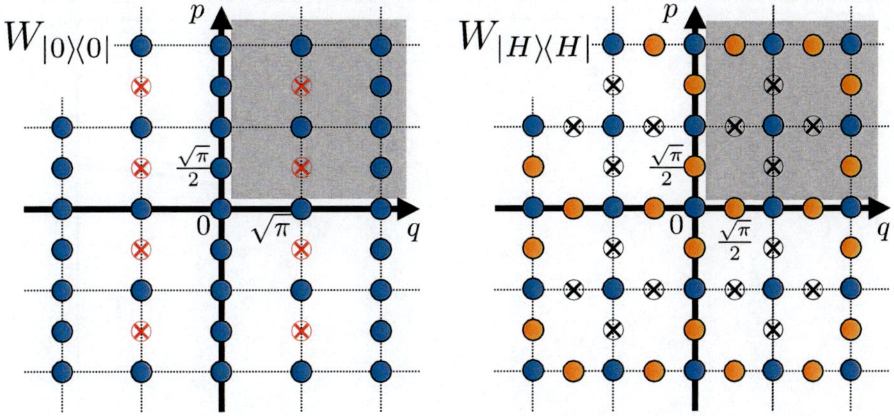

Fig. 5.7 The schematics of the Wigner functions of the ideal GKP states $|0\rangle$ on the left and $|H\rangle$ on the right, where each blue filled circle represents a positive delta function $\delta(\boldsymbol{r})$ with $\boldsymbol{r} := (q, p)$, each red circled X represents a negative delta function $-\delta(\boldsymbol{r})$, each yellow filled circle represents a weighted positive delta function $\frac{1}{\sqrt{2}}\delta(\boldsymbol{r})$, and each black circled X represents a weighted negative delta function $-\frac{1}{\sqrt{2}}\delta(\boldsymbol{r})$. These can be derived from Eq. (5.79) with the limit $\sigma_q^2, \sigma_p^2 \to 0$. The gray region shows the periodicity of these Wigner functions

the non-Gaussianity of the ideal GKP states exploiting the periodicity of the Wigner functions as shown in Fig. 5.7 and replacing the improper integral of $\mathcal{N}_L$ from $-\infty$ to ∞ with an integral over a period. More precisely, in place of $\mathcal{N}_L$, we define

$$\widetilde{\mathcal{N}}_L(\psi) := \ln\left(\frac{\iint_{\mathcal{I}} dq dp \, |W_\psi(q, p)|}{\iint_{\mathcal{I}} dq dp \, W_\psi(q, p)}\right), \tag{5.98}$$

where $\mathcal{I} := \left[0 + \epsilon, 2\sqrt{\pi} + \epsilon\right]$ for any fixed $\epsilon \in \left(0, \sqrt{\pi}/2\right)$ represents the period shown in Fig. 5.7. It is defined so that $\widetilde{\mathcal{N}}_L(\psi) = 0$ for a state ψ whose Wigner function is nonnegative. Then, by counting delta functions in Fig. 5.7, we have

$$\widetilde{\mathcal{N}}_L(|0\rangle\langle 0|) = \ln\frac{8}{4} = \ln 2, \tag{5.99}$$

$$\widetilde{\mathcal{N}}_L(|H\rangle\langle H|) = \ln\frac{4 + 8 \times \left(1/\sqrt{2}\right)}{4} = \ln(1 + \sqrt{2}). \tag{5.100}$$

Thus, we quantitatively compare the non-Gaussianity of $|H\rangle$ and $|0\rangle$ by

$$\widetilde{\mathcal{N}}_L(|H\rangle\langle H|) - \widetilde{\mathcal{N}}_L(|0\rangle\langle 0|) = \ln\frac{1 + \sqrt{2}}{2} > 0, \tag{5.101}$$

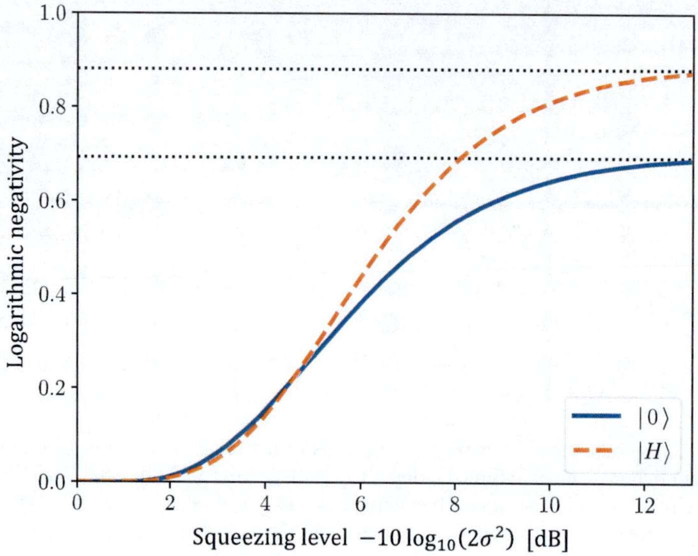

Fig. 5.8 The logarithmic negativities of the Wigner functions of $|0_{\sigma^2}\rangle$ (blue solid line) and $|H_{\sigma^2}\rangle \propto \left(\cos(\pi/8)|0_{\sigma^2}\rangle + \sin(\pi/8)|1_{\sigma^2}\rangle\right)$ (orange dashed line) with respect to the squeezing level $-10\log_{10}(2\sigma^2)$. The logarithmic negativity of $|0_{\sigma^2}\rangle$ approaches to $\ln(2) = 0.69\cdots$, and that of $|H_{\sigma^2}\rangle$ to $\ln(1+\sqrt{2}) = 0.88\cdots$, as expected from the calculations (5.99) and (5.100)

which implies that $|H\rangle$ has more non-Gaussianity than $|0\rangle$, and hence no Gaussian operation can deterministically transform $|0\rangle$ into $|H\rangle$.

In order to justify the use of $\widetilde{\mathcal{N}}_L$ instead of $\mathcal{N}_L$ for the ideal GKP states, we carried out a numerical simulation of the negativity $\mathcal{N}_L$ of approximate GKP states. More precisely, we performed the numerical integration using Mathematica 11.2.0 for the absolute values of the Wigner functions of $|0_{\sigma^2}\rangle$ and $|H_{\sigma^2}\rangle \propto (\cos(\pi/8)|0_{\sigma^2}\rangle + \sin(\pi/8)|1_{\sigma^2}\rangle)$ with the approximate GKP states $|j_{\sigma^2}\rangle$ defined in Eq. (5.72). Figure 5.8[9] plots the logarithmic negativities of the Wigner functions of $|0_{\sigma^2}\rangle$ and $|H_{\sigma^2}\rangle$ against the squeezing level $-10\log_{10}(2\sigma^2)$. The figure indicates that the logarithmic negativity of $|0_{\sigma^2}\rangle$ approaches to $\widetilde{\mathcal{N}}_L(|0\rangle\langle 0|) = \ln(2) = 0.69\cdots$ as $-10\log_{10}(2\sigma^2) \to \infty$ and that of $|H_{\sigma^2}\rangle$ to $\widetilde{\mathcal{N}}_L(|H\rangle\langle H|) = \ln(1+\sqrt{2}) = 0.88\cdots$, as expected from our arguments.

With $\widetilde{\mathcal{N}}_L$ defined, we can provide an upper bound on the conversions from the GKP $|0\rangle$s to the GKP $|H\rangle$s or the conversions in the reverse direction. In the same way as the additivity of the logarithmic negativity $\mathcal{N}_L$ [68, 69], $\widetilde{\mathcal{N}}_L$ is additive, i.e., $\widetilde{\mathcal{N}}_L(\psi^{\otimes n}) = n\,\widetilde{\mathcal{N}}_L(\psi)$. Although we have developed the protocol converting $|H\rangle^{\otimes 2}$ into $|0\rangle$ under Gaussian operations, the additivity of $\widetilde{\mathcal{N}}_L$ shows that $|H\rangle^{\otimes 2}$ cannot be transformed into $|0\rangle^{\otimes 3}$ under any Gaussian operations because of the inequality

[9] Modified from [45], Copyright 2020 by American Physical Society.

$$\widetilde{\mathcal{N}}_L(|H\rangle\langle H|^{\otimes 2}) - \widetilde{\mathcal{N}}_L(|0\rangle\langle 0|^{\otimes 3}) = \ln \frac{(1+\sqrt{2})^2}{2^3} < 0. \quad (5.102)$$

Whether the one-to-one transformation from $|H\rangle$ to $|0\rangle$ is possible or not is unknown. On the other hand, the conversion from $|0\rangle^{\otimes 2}$ to $|H\rangle$ is not prohibited in terms of the logarithmic negativity since

$$\widetilde{\mathcal{N}}_L(|0\rangle\langle 0|^{\otimes 2}) - \widetilde{\mathcal{N}}_L(|H\rangle\langle H|) = \ln \frac{2^2}{1+\sqrt{2}} > 0, \quad (5.103)$$

whereas it is still open whether a deterministic or exact Gaussian transformation from a constant number of $|0\rangle$s to $|H\rangle$ is possible or not. We here remark that the estimation on the required resource for the state conversion made by the logarithmic negativity is in general very loose; the state with larger logarithmic negativity does not necessarily mean the convertibility to the state with smaller logarithmic negativity [68, 69].

Note that the measure $\widetilde{\mathcal{N}}_L$ was also introduced in Ref. [97], which is done independently. In contrast to Ref. [97], we justified the use of $\widetilde{\mathcal{N}}_L$ by numerically calculating the original logarithmic negativity $\mathcal{N}_L$ for the approximate GKP states and then taking the limit of the infinite squeezing level as shown in Fig. 5.8. Furthermore, we explicitly show an example for the transformation under Gaussian operations in Fig. 5.6, which supports the usefulness of the resource theory.

To obtain further insight on the convertibility between GKP states, more sophisticated resource-theoretic analyses may be needed. For example, an interesting possibility is that Gaussian operations combined with post-selection may increase the negativity of a GKP state with nonzero probability. Such a possibility is not covered by the analysis using the logarithmic negativity. Our resource-theoretical arguments and methods are thus starting points for tackling questions on the GKP-state conversion.

5.4.3 *Feasibility of Preparing a GKP Magic State*

Since Gaussian operations combined with a GKP magic state $|H\rangle$ can be advantageous in implementing QC over those with $|0\rangle$ of GKP qubits, we here discuss possible protocols for directly preparing $|H\rangle$. Note that choosing $|H\rangle$ among GKP-encoded magic states may not just be a matter of convention; the $|H\rangle$ state has the symmetry of $\frac{\pi}{2}$-rotation in the phase space as seen in Fig. 5.7 and thus lies in an eigenspace of the photon number operator modulo four. Some proposals exploit this symmetry for the preparation [1]. Our following discussion is based on the proposals for the quantum optical implementation of approximate GKP qubits [39, 50, 52–54, 59, 71–74, 83–85], while there also exist other proposals and experimental demonstrations of generating approximate GKP states in various systems [56, 62, 64, 65, 75–82]. Note that architectures such as superconducting cavities [65] and trapped-ion mechanical oscillators [64] are also promising candidates to realize the

GKP code, but we here focus on quantum optical implementations since Gaussian operations are not necessarily easier to implement than non-Gaussian operations on the superconducting cavities and the trapped-ion mechanical oscillators. We remark that these existing proposals mostly focus on preparing $|0\rangle$ or $|1\rangle$ of the GKP code.

Some of the proposals, such as those in Refs. [50, 83, 85], may not be suitable for the direct preparation of $|H\rangle$. References [50, 85] considers using the cross-Kerr non-linearity to couple two optical modes initially prepared in a coherent state and a squeezed coherent state, respectively, followed by performing homodyne measurement of the mode initialized as the coherent state, which results in generating approximate GKP-Pauli eigenstate $|0\rangle$ or $|1\rangle$. In this scheme, $|H\rangle$ cannot be directly prepared as long as Gaussian states are fed into the cross-Kerr interaction followed by the homodyne detection. Reference [83] considers using the optical displacement and the controlled-π rotation controlled by the state of the two-level atom. This setup cannot directly prepare the $|H\rangle$ state since the controlled-π rotation cannot add relative phases between superposed peaks. As for other proposals, Ref. [72] analyzes optimization of parametrized non-Gaussian optical circuits by machine learning, and Ref. [73] uses time-frequency degrees of freedom. These proposals do not fit our current settings for preparing the quantum optical GKP qubits where Gaussian operations are easy compared to non-Gaussian operations, while they are also interesting research directions.

There are, however, some existing proposals that can be used for preparing $|H\rangle$. One approach is to breed the peaks of the approximate GKP states by the controlled-displacement operation using the interaction with the discrete-variable system in the cavity QED setups [54, 59, 82] or by the interference between two premature approximate GKP states in the all-optical setups [52, 53, 71]. Another approach is to use linear optical circuits followed by photon-number-resolving (PNR) detectors and post-select certain outcome patterns [39, 74, 84].

For the former approach, $|H\rangle$ can be prepared if the circuit given in Fig. 5.9 can be implemented by the same experimentally allowed operations that prepare $|0\rangle$. Reference [53] develops a refinement of the proposals in Refs. [52, 71] of the all-optical interference-based breeding protocol. The method proposed in Ref. [53] can realize the circuit equivalent to that in Fig. 5.9 by post-selecting the homodyne outcome and adjusting the relative phase between $|0\rangle$ and $|1\rangle$. (In principle, an arbitrary superposition of $|0\rangle$ and $|1\rangle$ can be prepared with the method in Ref. [53].) Reference [82], which uses the recursive interaction between a discrete-variable system and a continuous-variable system, can also realize the circuit equivalent to that in Fig. 5.9 by adjusting the interaction terms. (The explicit methodology is given in the literature.) Thus in these protocols, the technological requirements for preparing $|0\rangle$ and $|H\rangle$ are at the same level. Note that if we are allowed to use an interaction between a qubit and a continuous-variable system beyond the controlled-displacement gate, an additional controlled-Fourier operation between the qubit and the continuous-variable system can also prepare $|H\rangle$ from $|0\rangle$ as already shown in Ref. [1].

For the latter approach, generating the non-Gaussian states with linear optical circuits followed by the PNR detectors as proposed in Refs. [39, 74, 84] can implement $|0\rangle$ and $|H\rangle$ on an equal footing with almost the same resource requirements,

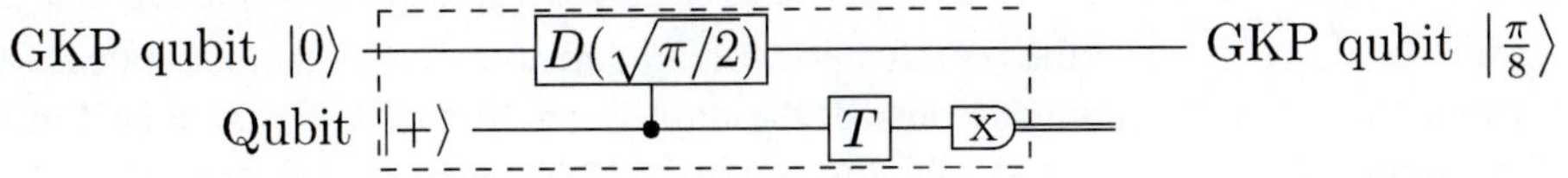

Fig. 5.9 A quantum circuit for preparing a GKP $\frac{\pi}{8}$-phase state $|\frac{\pi}{8}\rangle$ from an input GKP $|0\rangle$ state using the breeding-like (phase-estimation-like) circuit (Modified from [45], Copyright 2020 by American Physical Society.)

as pointed out in Ref. [39]. Furthermore, preparing only one type of GKP state in this protocol may be desired since optical circuits and PNR detectors to generate the GKP state need to be finely tuned to keep the fidelity high. These protocols indicate that it is feasible to prepare $|H\rangle$ of the GKP qubits with a technological requirement comparable to preparing $|0\rangle$ in some cases.

5.4.4 *Discussion*

We have proposed the implementation of the universal continuous-variable QC based on the preparation of $|H\rangle$ rather than $|0\rangle$ of the GKP code combined with the Gaussian operations. Our main contribution is the fundamental cost reduction of the non-Gaussian resources compared to the existing proposal [43] by the direct preparation of $|H\rangle$. This can avoid the costly distillation in [43] for converting $|0\rangle$ states into a $|H\rangle$ state, and achieve a constant overhead in converting $|H\rangle$ states into a $|0\rangle$ state. This cost reduction holds under the condition that the preparation of the non-Gaussian states such as the GKP $|0\rangle$ and $|H\rangle$ states are equally costly compared to Gaussian operations, which is true in the quantum optical system. It holds neither for qubit error-correcting codes in which the preparation of the magic state is much more costly than that of the Pauli eigenstate nor for the continuous-variable systems other than the quantum optical system in which Gaussian operations are not necessarily easy compared to the non-Gaussian operations. In addition to the development of the cost-reduced protocol, we have introduced the measure $\widetilde{\mathcal{N}}_L$ to quantify the non-Gaussianity of the ideal GKP states and addressed the fundamental limitation on the convertibility between GKP $|H\rangle$ and $|0\rangle$ under Gaussian operations. Our findings of the usefulness of the resource theory of non-Gaussianity in the continuous-variable QC may open up new directions for the applications of the resource theory in continuous-variable systems. We have also discussed two promising protocols for the direct preparation of $|H\rangle$ in the quantum optical system; one is based on the breeding of peaks of approximate GKP states (Refs. [53, 82]), and the other is on the photon-number-resolving measurement (Refs. [39, 74, 84]). We point out that not much attention has been paid to the direct preparations of the GKP magic state $|H\rangle$. Our results thus put forward an argument on which of the GKP-encoded state should be used for the initialization to implement the fault-tolerant optical QC, while more concrete comparisons for the implementation cost may require further assumptions on advances in quantum optical technologies and hence are left for future work.

5.5 Conclusion for This Chapter

In this chapter, we focus on the GKP code, which encodes a digitized degree of freedom into a continuou-variable system. The GKP code can correct displacement errors in the continuous-variable system up to half the lattice spacing of the GKP-encoded state. The ideal GKP code, however, assumes the use of an unphysical state and should thus be regarded as the limit of its approximations. There are several conventional approximations of the GKP states, which are thought to be equivalent at least in the limit of good approximation, but the lack of explicit correspondence prevents the accurate comparison between researches based on the different conventions. In Sect. 5.3, we showed that these conventional (symmetric) approximations of the GKP states are equivalent by showing the exact correspondence between approximation parameters. This result motivates us to introduce the standard form of approximate GKP codes, and for the standard form, we derived their Wigner functions, the inner products, and the average photon numbers. Our results bridge the gap for the past studies based on different approximations and may be useful to analyze the recent and near-future experimental realizations of the approximate GKP-encoded states.

There are several reasons why the GKP code is suitable for the optical fault-tolerant continuous-variable QC. The foremost reason is that Gaussian operations, which can be reliably implemented in quantum optical systems, are sufficient, in addition to the preparation of the GKP-encoded states, to achieve the fault-tolerant universal QC. In Sect. 5.4, we showed that, in addition to Gaussian operations, the ability to prepare only one type of the GKP-encoded magic state is sufficient to realize the fault-tolerant universality in a cost-efficient way. The cost-efficiency is achieved by the deterministic state conversion from the GKP magic states to the GKP Pauli state with Gaussian operations, while the known state conversion from the GKP Pauli states to the GKP magic state is probabilistic. To analyze the state convertibility between the GKP-encoded states under Gaussian operations, we apply the resource theory of non-Gaussianity. Our argument of cost efficiency is implicitly based on the fact that the required cost for generating the GKP magic state is comparable to that for generating the GKP Pauli state, which is not true for the majority of the qubit-based quantum error-correcting codes. To justify this, we proposed possible ways to directly prepare the GKP magic state with slight modifications to the existing proposals for the experimental GKP Pauli state realization.

With the results in Sects. 5.3 and 5.4 combined, the (approximate) GKP code is shown to be a powerful tool to realize the optical fault-tolerant continuous-variable

QC [98, 99]. Our results thus show rich potentials of the continuous-variable system for realizing the fault-tolerant universal QC by encoding the digitized quantum information into the continuous-variable system such as the GKP code.

References

1. Gottesman D, Kitaev A, Preskill J (2001) Encoding a qubit in an oscillator. Phys Rev 64(1):012310
2. Nielsen MA, Chuang IL (2010) Quantum computation and quantum information, 10th edn. Cambridge University Press. https://doi.org/10.1017/CBO9780511976667.
3. Eisert J, Plenio MB (2003) Introduction to the basics of entanglement theory in continuous-variable systems. Int J Quantum Inf 1(04):479–506
4. Wilde MM, Qi H (2018) Energy-constrained private and quantum capacities of quantum channels. IEEE Trans Inf Theory 64(12):7802–7827. https://doi.org/10.1109/TIT.2018.2854766.
5. Harrow AW, Montanaro A (2017) Quantum computational supremacy. Nature 549:203–209. https://doi.org/10.1038/nature23458, https://www.nature.com/articles/nature23458
6. Arute F et al (2019) Quantum supremacy using a programmable superconducting processor. Nature 574(7779):505–510. issn: 1476-4687. https://doi.org/10.1038/s41586-019-1666-5
7. Bravyi S, Gosset D, König R (2018) Quantum advantage with shallow circuits. Science 362(6412):308–311
8. Broadbent A, Fitzsimons J, Kashefi E (2009) Universal blind quantum computation. In: 2009 50th annual IEEE symposium on foundations of computer science, pp 517–526. https://doi.org/10.1109/FOCS.2009.36, https://ieeexplore.ieee.org/document/5438603
9. Broadbent A, Fitzsimons J, Kashefi E (2010) Measurement-based and universal blind quantum computation. In: International school on formal methods for the design of computer, communication and software systems. Springer, Berlin, pp 43–86
10. Wendin G (2017) Quantum information processing with superconducting circuits: a review. Rep Prog Phys 80(10):106001. https://doi.org/10.1088/1361-6633/aa7e1a, http://iopscience.iop.org/article/10.1088/1361-6633/aa7e1a/
11. Krantz P et al (2019) A quantum engineer's guide to superconducting qubits. Appl Phys Rev 6(2):021318. https://doi.org/10.1063/1.5089550, https://aip.scitation.org/doi/10.1063/1.5089550
12. Häffner H, Roos CF, Blatt R (2008) Quantum computing with trapped ions. Phys Rep 469(4):155–203. https://doi.org/10.1016/j.physrep.2008.09.003, http://www.sciencedirect.com/science/article/pii/S0370157308003463
13. Bruzewicz CD et al (2019) Trapped-ion quantum computing: progress and challenges. Appl Phys Rev 6(2):021314. https://doi.org/10.1063/1.5088164
14. Yoshikawa J-I et al (2016) Invited article: generation of one-million-mode continuous-variable cluster state by unlimited time-domain multiplexing. APL Photonics 1(6):060801. https://aip.scitation.org/doi/full/10.1063/1.4962732
15. Lloyd S, Slotine J-JE (1998) Analog quantum error correction. Phys Rev Lett 80(18):4088
16. Braunstein SL (1998) Error correction for continuous quantum variables. In: Quantum information with continuous variables. Springer, Berlin, pp 19–29
17. Menicucci NC (2014) Fault-tolerant measurement-based quantum computing with continuous-variable cluster states. Phys Rev Lett 112(12):120504
18. Ketterer A et al (2016) Quantum information processing in phase space: a modular variables approach. Phys Rev 94(2):022325
19. Cochrane PT, Milburn GJ, Munro WJ (1999) Macroscopically distinct quantum-superposition states as a bosonic code for amplitude damping. Phys Rev 59(4):2631
20. Niset J, Andersen UL, Cerf NJ (2008) Experimentally feasible quantum erasure-correcting code for continuous variables. Phys Rev Lett 101(13):130503

21. Leghtas Z et al (2013) Hardware-efficient autonomous quantum memory protection. Phys Rev Lett 111(12):120501
22. Felipe L, Renes JM, Scholz VB (2016) Coherent state constellations for Bosonic Gaussian channels. In: IEEE international symposium on information theory (ISIT). IEEE, pp 2499–2503
23. Lacerda F, Renes JM, Scholz VB (2017) Coherent-state constellations and polar codes for thermal Gaussian channels. Phys Rev 95(6):062343
24. Chuang IL, Leung DW, Yamamoto Y (1997) Bosonic quantum codes for amplitude damping. Phys Rev 56(2):1114
25. Knill E, Laflamme R, Milburn GJ (2001) A scheme for efficient quantum computation with linear optics. Nature 409(6816):46
26. Ralph TC, Hayes AJF, Gilchrist A (2005) Loss-tolerant optical qubits. Phys Rev Lett 95(10):100501
27. Wasilewski W, Banaszek K (2007) Protecting an optical qubit against photon loss. Phys Rev 75(4):042316
28. Bergmann M, van Loock P (2016) Quantum error correction against photon loss using NOON states. Phys Rev 94(1):012311
29. Michael MH et al (2016) New class of quantum error-correcting codes for a bosonic mode. Phys Rev X 6(3):031006
30. Niu MY, Chuang IL, Shapiro JH (2018) Hardware-efficient bosonic quantum error-correcting codes based on symmetry operators. Phys Rev A 97(3):032323
31. Albert VV et al (2018) Performance and structure of single-mode bosonic codes. Phys Rev A 97(3):032346
32. Douce T et al (2017) Continuous-variable instantaneous quantum computing is hard to sample. Phys Rev Lett 118:070503. https://doi.org/10.1103/PhysRevLett.118.070503
33. Fukui K, Tomita A, Okamoto A (2017) Analog quantum error correction with encoding a qubit into an oscillator. Phys Rev Lett 119(18):180507
34. Fukui K et al (2018) High-threshold fault-tolerant quantum computation with analog quantum error correction. Phys Rev X 8:021054. https://doi.org/10.1103/PhysRevX.8.021054
35. Vuillot C et al (2019) Quantum error correction with the toric Gottesman-Kitaev-Preskill code. Phys Rev A 99(3):032344
36. Walshe BW et al (2019) Robust fault tolerance for continuous-variable cluster states with excess antisqueezing. Phys Rev A 100 :010301. https://doi.org/10.1103/PhysRevA.100.010301
37. Wang Y (2019) Quantum error correction with the GKP code and concatenation with stabilizer codes. arXiv: 1908.00147 [quant-ph]
38. Noh K, Chamberland C (2020) Fault-tolerant bosonic quantum error correction with the surface-Gottesman-Kitaev-Preskill code. Phys Rev A 101(1):012316
39. Tzitrin I et al (2020) Progress towards practical qubit computation using approximate Gottesman-Kitaev-Preskill codes. Phys Rev A 101:032315. https://doi.org/10.1103/PhysRevA.101.032315
40. Hänggli L, Heinze M, König R (2020) Enhanced noise resilience of the surface-Gottesman-Kitaev-Preskill code via designed bias. Phys Rev A 102:052408. https://doi.org/10.1103/PhysRevA.102.052408
41. Harrington J, Preskill J (2001) Achievable rates for the Gaussian quantum channel. Phys Rev A 64(6):062301
42. Noh K, Albert VV, Jiang L (2018) Quantum capacity bounds of Gaussian thermal loss channels and achievable rates with Gottesman-Kitaev-Preskill codes. IEEE Trans Inf Theory 65(4):2563–2582
43. Baragiola BQ et al (2019) All-Gaussian universality and fault tolerance with the Gottesman-Kitaev-Preskill code. Phys Rev Lett 123(20):200502
44. Matsuura T, Yamasaki H, Koashi M (2020) Equivalence of approximate Gottesman-Kitaev-Preskill codes. Phys Rev A 102:032408. https://doi.org/10.1103/PhysRevA.102.032408,
45. Yamasaki H, Matsuura T, Koashi M (2020) Cost-reduced all- Gaussian universality with the Gottesman-Kitaev-Preskill code: resource-theoretic approach to cost analysis. Phys Rev Res 2:023270. https://doi.org/10.1103/PhysRevResearch.2.023270

46. Kitaev AY et al (2002) Classical and quantum computation, vol. 47. Graduate Studies in Mathematics. American Mathematical Society
47. Dawson CM, Nielsen MA (2005) The solovay-kitaev algorithm. arXiv: 0505030 [quant-ph]
48. Gottesman D (1998) The Heisenberg representation of quantum computers. arXiv: 9807006 [quant-ph]
49. Aaronson S, Gottesman D (2004) Improved simulation of stabilizer circuits. Phys Rev A 70:052328. https://doi.org/10.1103/PhysRevA.70.052328. url: https://link.aps.org/doi/10.1103/PhysRevA.70.052328
50. Pirandola S et al (2004) Constructing finite-dimensional codes with optical continuous variables. EPL (Europhysics Letters) 68(3):323. https://iopscience.iop.org/article/10.1209/epl/i2004-10203-9/meta
51. Glancy S, Knill E (2006) Error analysis for encoding a qubit in an oscillator. Phys Rev A 73(1):012325
52. Vasconcelos HM, Sanz L, Glancy S (2010) All-optical generation of states for Encoding a qubit in an oscillator. Opt Lett 35(19):3261–3263
53. Weigand DJ, Terhal BM (2018) Generating grid states from Schrödinger-cat states without postselection. Phys Rev A 97(2):022341
54. Motes KR et al (2017) Encoding qubits into oscillators with atomic ensembles and squeezed light. Phys Rev A 95(5):053819
55. Galetti D, Marchiolli MA (1996) Discrete coherent states and probability distributions in finite-dimensional spaces. Ann Phys 249(2):454–480
56. Terhal BM, Weigand D (2016) Encoding a qubit into a cavity mode in circuit QED using phase estimation. Phys Rev A 93(1):012315
57. Duivenvoorden K, Terhal BM, Weigand D (2017) Single-mode displacement sensor. Phys Rev A 95(1):012305
58. Mumford D, Musili C (2007) Tata lectures on theta. I (Modern Birkhäuser classics). Birkhäuser Boston Incorporated
59. Travaglione BC, Milburn GJ (2002) Preparing encoded states in an oscillator. Phys Rev A 66(5):052322
60. Pantaleoni G, Baragiola BQ, Menicucci NC (2020) Modular Bosonic subsystem codes. Phys Rev Lett 125: 040501. https://doi.org/10.1103/PhysRevLett.125.040501, https://link.aps.org/doi/10.1103/PhysRevLett.125.040501
61. Deconinck B et al (2004) Computing Riemann theta functions. Math Comput 73(247):1417–1442
62. Shi Y, Chamberland C, Cross A (2019) Fault-tolerant preparation of approximate GKP states. New J Phys 21(9):093007
63. Berndt B, Kim B (2011) Asymptotic expansions of certain partial theta functions. Proc Am Math Soc 139(11):3779–3788
64. Flühmann C et al (2019) Encoding a qubit in a trapped-ion mechanical oscillator. Nature 566(7745):513
65. Campagne-Ibarcq P et al (2020) Quantum error correction of a qubit encoded in grid states of an oscillator. Nature 584(7821):368–372
66. Holevo AS, Sohma M, Hirota O (1999) Capacity of quantum Gaussian channels. Phys Rev A 59(3):1820
67. Bravyi S, Kitaev A (2005) Universal quantum computation with ideal Clifford gates and noisy ancillas. Phys Rev A 71:022316. https://doi.org/10.1103/PhysRevA.71.022316
68. Takagi R, Zhuang Q (2018) Convex resource theory of non-Gaussianity. Phys Rev A 97:062337. https://doi.org/10.1103/PhysRevA.97.062337
69. Albarelli F et al (2018) Resource theory of quantum non-Gaussianity and Wigner negativity. Phys Rev A 98:052350. https://doi.org/10.1103/PhysRevA.98.052350, https://link.aps.org/doi/10.1103/PhysRevA.98.052350
70. Chitambar E, Gour G (2019) Quantum resource theories. Rev Mod Phys 91:025001. https://doi.org/10.1103/RevModPhys.91.025001, https://link.aps.org/doi/10.1103/RevModPhys.91.025001

71. Jean Etesse et al. "Proposal for a loophole-free violation of Bell's inequalities with a set of single photons and homodyne measurements". In: New J. Phys. 16.5 (2014), p. 053001. url: https://iopscience.iop.org/article/10.1088/1367-2630/16/5/053001/meta
72. Arrazola JM et al (2018) Machine learning method for state preparation and gate synthesis on photonic quantum computers. Quantum Sci Tech (2018). https://iopscience.iop.org/article/10.1088/2058-9565/aaf59e/pdf
73. Fabre N et al (2020) Generation of a time-frequency grid state with integrated biphoton frequency combs. Phys Rev A 102:012607. https://doi.org/10.1103/PhysRevA.102.012607, https://link.aps.org/doi/10.1103/PhysRevA.102.012607
74. Eaton M, Nehra R, Pfister O (2019) Non-Gaussian and Gottesman-Kitaev-Preskill state preparation by photon catalysis. New J Phys 21(11):113034
75. Pirandola S et al (2006) Continuous variable encoding by ponderomotive interaction. Euro Phys J D 37(2):283–290. https://link.springer.com/article/10.1140/epjd/e2005-00306-3
76. Pirandola S et al (2006) Generating continuous variable quantum codewords in the near-field atomic lithography. J Phys B 39(4):997. https://doi.org/10.1088/0953-4075/39/4/023
77. Brooks P, Kitaev A, Preskill J (2013) Protected gates for superconducting qubits. Phys Rev A 87(5):052306. https://journals.aps.org/pra/abstract/10.1103/PhysRevA.87.052306
78. Flühmann C et al (2018) Sequential modular position and momentum measurements of a trapped ion mechanical oscillator. Phys Rev X 8(2):021001. https://journals.aps.org/prx/abstract/10.1103/PhysRevX.8.021001
79. Lin C-Y, Su W-C, Wu S-T (2020) Encoding qubits into harmonicoscillator modes via quantum walks in phase space. Quantum Inf Process 19(8):272. issn: 1573-1332. https://doi.org/10.1007/s11128-020-02775-6
80. Weigand DJ, Terhal BM (2020) Realizing modular quadrature measurements via a tunable photon-pressure coupling in circuit QED. Phys Rev A 101:053840. https://doi.org/10.1103/PhysRevA.101.053840
81. Le DT et al (2019) Doubly nonlinear superconducting qubit. Phys Rev A 100:062321. https://doi.org/10.1103/PhysRevA.100.062321, https://link.aps.org/doi/10.1103/PhysRevA.100.062321
82. Hastrup J et al (2021) Measurement-free preparation of grid states. NPJ Quantum Inf 7(1):17. issn: 2056-6387. https://doi.org/10.1038/s41534-020-00353-3
83. Hastrup J, Andersen UL (2021) Generation of optical Gottesman-Kitaev-Preskill states with cavity QED. arXiv: 2104.07981 [quant-ph]
84. Fukui K et al (2021) Efficient backcasting search for optical quantum state synthesis. arXiv: 2109.01444 [quant-ph]
85. Fukui K et al (2021) Generating Gottesman-Kitaev-Preskill qubit using a cross- Kerr interaction between a squeezed light and Fock states in optics. arXiv: 2109.04801 [quant-ph]
86. Niset J, Fiurášek J, Cerf NJ (2009) No-Go theorem for Gaussian quantum error correction. Phys Rev Lett 102:120501. https://doi.org/10.1103/PhysRevLett.102.120501, https://link.aps.org/doi/10.1103/PhysRevLett.102.120501
87. Reichardt BW (2005) Quantum universality from magic states distillation applied to CSS codes. Quantum Inf Process 4(3):251–264. issn: 1573-1332. https://doi.org/10.1007/s11128-005-7654-8
88. Bravyi S, Haah J (2012) Magic-state distillation with low overhead. Phys Rev A 86:052329. https://doi.org/10.1103/PhysRevA.86.052329, https://link.aps.org/doi/10.1103/PhysRevA.86.052329
89. Jones C (2013) Multilevel distillation of magic states for quantum computing. Phys Rev A 87:042305. https://doi.org/10.1103/PhysRevA.87.042305, https://link.aps.org/doi/10.1103/PhysRevA.87.042305
90. Haah J et al (2017) Magic state distillation with low space overhead and optimal asymptotic input count. Quantum 1:31. https://doi.org/10.22331/q-2017-10-03-31
91. Lami L et al (2018) Gaussian quantum resource theories. Phys Rev A 98:022335. https://doi.org/10.1103/PhysRevA.98.022335

92. Zhuang Q, Shor PW, Shapiro JH (2018) Resource theory of non-Gaussian operations. Phys Rev A 97:052317. https://doi.org/10.1103/PhysRevA.97.052317, https://link.aps.org/doi/10.1103/PhysRevA.97.052317
93. Veitch V et al (2014) The resource theory of stabilizer quantum computation. New J Phys 16(1):013009. https://doi.org/10.1088/1367-2630/16/1/013009
94. Howard M, Campbell E (2014) Application of a resource theory for magic states to fault-tolerant quantum computing. Phys Rev Lett 118:090501. https://doi.org/10.1103/PhysRevLett.118.090501, https://link.aps.org/doi/10.1103/PhysRevLett.118.090501
95. Hudson RL (1974) When is the wigner quasi-probability density non-negative? Rep Math Phys 6(2):249–252. issn: 0034-4877. https://www.sciencedirect.com/science/article/pii/003448777490007X
96. Soto F, Claverie P (1983) When is the Wigner function of multidimensional systems nonnegative? J Math Phys 24(1):97–100. https://doi.org/10.1063/1.525607
97. García-Álvarez L, Ferraro A, Ferrini G (2021) From the bloch sphere to phase- space representations with the Gottesman-Kitaev-Preskill encoding. In: Takagi T et al (ed) International symposium on mathematics, quantum theory, and cryptography. Springer, Singapore, pp 79–92. isbn: 978-981-15-5191-8
98. Eli Bourassa J et al (2021) Blueprint for a scalable photonic fault-tolerant quantum computer. Quantum 5:392. issn: 2521-327X. https://doi.org/10.22331/q-2021-02-04-392
99. Tzitrin I et al (2021) Fault-tolerant quantum computation with static linear optics. PRX Quantum 2:040353. https://doi.org/10.1103/PRXQuantum.2.040353, https://link.aps.org/doi/10.1103/PRXQuantum.2.040353

Chapter 6
Conclusion

Abstract This chapter concludes the thesis. Special emphasis is on the achievements made by this thesis and possible impacts on the fields.

Keywords Continuous-variable system · Quantum key distribution · Quantum computation

6.1 Conclusion

For experimental implementations of quantum information processing, a promising candidate for the hardware is a continuous-variable system, a quantum optical system as a prominent example. In this thesis, we studied digital quantum information processing with continuous-variable systems. More specifically, we studied continuous-variable quantum key distribution (QKD) and continuous-variable quantum computation (QC) based on digitized information processing.

For continuous-variable QKD, we developed a binary-modulation protocol that is adapted to digital signal processing and proved its composable security against general attacks in the case of a finite number of communication rounds. This complete security proof was obtained by the newly developed fidelity estimation to an arbitrary coherent state using the heterodyne measurement and the classical post-processing. Using the fidelity as the measure of disturbance, we obtained an upper bound on the number of phase errors that reflect the information leakage to the eavesdropper by the heuristic operator inequality. The finite-size security against general attacks for discrete-modulation continuous-variable QKD protocols had been an open problem; our result partially solves the problem and can be a milestone for the ultimate goal. The obtained key rate with our security proof did not, however, achieve the asymptotically optimal scaling rate estimated in previous studies. Therefore, we further refined the proof based on the reverse reconciliation, the frequently used technique in the field of the continuous-variable QKD, with a slight modification to the classi-

T. Matsuura, *Digital Quantum Information Processing with Continuous-Variable Systems*, Springer Theses, https://doi.org/10.1007/978-981-19-8288-0_6

cal post-processing in the original protocol while keeping the original experimental setups unchanged. As a result, under the pure-loss channel, the protocol asymptotically achieves an almost optimal key-rate scaling against the transmissivity. Thus, the developed protocol turns out to have fairly good performance for quantum channels that are close to the pure loss. Based on the results in this thesis, the next step is to generalize the security proof to more general discrete-modulation continuous-variable QKD protocols and obtain higher tolerance against excess noises, which has practical as well as theoretical importance.

For continuous-variable QC, we gave several results on the GKP code, which encodes a qudit into a continuous-variable system and has many desirable properties for optical QC. The GKP code can be implemented in a physical system only approximately, and thus several approximations of the GKP code have been widely used. We showed that these conventional approximations are equivalent by showing the explicit correspondence between the approximation parameters. This result bridges the gap of previous studies that were based on different approximations. Furthermore, we constructed a resource-efficient protocol to realize the universal QC, requiring the preparation of only one type of the GKP-encoded magic state and Gaussian operations. We showed the deterministic GKP state conversion from the two GKP-encoded magic states to the GKP-encoded Pauli state while only the probabilistic conversion is known for the converse direction. We deepened the analysis for the convertibility of the GKP-encoded states using the resource theory of non-Gaussianity. Our resource estimation implicitly assumed that the preparation of the GKP magic states was as costly as the preparation of the GKP Pauli states. We justified this by showing that the slight modifications to some of the existing proposals for implementing GKP Pauli states can realize the GKP magic state preparation. Notably, for the proposals of using parametrized linear optical circuits followed by photon-number-resolving detectors to implement the GKP-encoded states, it is already known that the GKP magic state can be implemented on equal footing with the GKP Pauli state. Our results thus show the rich potentials of the GKP code for realizing fault-tolerant continuous-variable QC and may lead to mitigating the implementation cost of the optical fault-tolerant continuous-variable QC.

Throughout the thesis, we have developed information processing of quantum and classical digital information encoded in continuous-variable quantum systems. Thus, the contributions of this thesis are to elaborate the theory and enhance the experimental feasibility for digital information processing with the continuous-variable system, which opens up new possibilities for this field.

Appendix

Abstract In this chapter, the grid representation referred to in Chap. 5 is explained, and the proofs of propositions and lemmas in Chap. 5 is proved. Furthermore, alternative expressions of Wigner functions, inner products, normalization constants, and average photon numbers derived in Chap. 5 are given.

Keywords Wave function, Wigner function, Inner product, Normalization constant, Average photon number

A.1 The Grid Representation

The grid representation appeared in the paper by Zak [13], and was later elaborated upon [9, 14] and used in the context of quantum information theory [3, 8, 10, 11]. Here we make a brief review. Let $(u, v) \in [0, 1) \times [0, 1)$, and $\mathcal{V}(u, v) := V\left((2\pi v/(\alpha_d d), \alpha_d du)^\top\right)$. Then, $e^{-\pi it}\mathcal{V}$ forms a Heisenberg group $e^{-\pi it}\mathcal{V}(u, v) \cdot e^{-\pi it'}\mathcal{V}(u', v') = e^{-\pi i(t+t'+uv'-u'v)}\mathcal{V}(u+u', v+v')$. Define $|u, v\rangle_{\text{grid}}$ as

$$|u, v\rangle_{\text{grid}} := \mathcal{V}(u, v)\left|0^{(\text{ideal})}\right\rangle \tag{A.1}$$

$$= e^{-\pi iuv}Z(2\pi v/(\alpha_d d))X(\alpha_d du)\left|0^{(\text{ideal})}\right\rangle. \tag{A.2}$$

In Refs. [8, 10, 11], $|u, v\rangle_{\text{grid}}$ with $d = 1$ is called the "shifted grid state". The generalized "shifted grid state" $|u, v\rangle_{\text{grid}}$ with arbitrary d satisfies an orthogonality and completeness relation in the following sense [3, 8]:

T. Matsuura, *Digital Quantum Information Processing with Continuous-Variable Systems*, Springer Theses, https://doi.org/10.1007/978-981-19-8288-0

$$\langle u, v|u', v'\rangle_{\text{grid}} = \delta(u-u')\delta(v-v'),$$

$$\int_0^1 du \int_0^1 dv \; |u, v\rangle\langle u, v|_{\text{grid}} = I.$$

The "wave function" $\phi_f(u, v)$ of a state $|f\rangle$ with respect to the "shifted grid states", i.e., the grid representation of $|f\rangle$, is defined as $\phi_f(u, v) := \langle u, v|f\rangle$, which satisfies

$$\int_0^1 du \int_0^1 dv \left|\phi_f(u, v)\right|^2 = 1. \tag{A.3}$$

The "wave function" of the ideal GKP logical basis state $\left|j^{(\text{ideal})}\right\rangle$ can be regarded as a Dirac delta function centered at $(j/d, 0)$, which does not satisfy Eq. (A.3) and therefore, cannot be regarded as a physical state. However, functions satisfying Eq. (A.3) and localized at $(j/d, 0)$ are well-defined approximate logical basis states.

Given the position representation $\psi_f(q) := \langle q|f\rangle$ of a (pure) state $|f\rangle$, its grid representation $\phi_f(u, v)$ can be given by

$$\phi_f(u, v) := \langle u, v|f\rangle \tag{A.4}$$

$$= \int dq \; \langle u, v|_{\text{grid}} |q\rangle\langle q|_{\hat{q}} |f\rangle \tag{A.5}$$

$$= \sqrt{\alpha_d d} \sum_{s\in\mathbb{Z}} e^{-2\pi i v\left(s+\frac{u}{2}\right)} \psi_f\left(\alpha_d d(u+s)\right). \tag{A.6}$$

Using the last equality, we can expand the domain $[0, 1) \times [0, 1)$ of the "wave function" of the grid representation ϕ to $\mathbb{R}^2$. This redefined "wave function" $\phi : \mathbb{R}^2 \to \mathbb{C}$ satisfies Eq. (A.3) and the following:

$$\forall (n_1, n_2)^\top \in \mathbb{Z}^2, \qquad \phi(u+n_1, v+n_2) = e^{-\pi i(n_1 n_2 + u n_2 - v n_1)}\phi(u, v), \tag{A.7}$$

which can be confirmed from Eq. (A.6). The functions $\phi : \mathbb{R}^2 \to \mathbb{C}$ that satisfy Eqs. (A.3) and (A.7) form a representation space of the Heisenberg group called $L^2(\mathbb{R}^2/\mathbb{Z}^2)$ [15], where the action of the group element $\text{Op}(\cdot)$ on ϕ is given by

$$\text{Op}(e^{-\pi i}\mathcal{V}(u, v))\phi_f(x, y) := \langle x, y|_{\text{grid}} \, e^{-\pi i t}\mathcal{V}(u, v)\,|f\rangle \tag{A.8}$$

$$= e^{-\pi i(t+xv-yu)}\phi_f(x-u, y-v). \tag{A.9}$$

The formulation can easily be generalized to the g-mode case by considering the representation space $L^2(\mathbb{R}^{2g}/\mathbb{Z}^{2g})$ [15].

A.2 Proofs of the Propositions and Lemmas in Sect. 5.3

A.2.1 Proof of Proposition 5.3.2 and Lemma 5.3.3

First, we derive Eqs. (5.43), (5.45), and (5.47) in Proposition 5.3.2. In the main text, α is fixed to α_d for $\left|j^{(\mathrm{ideal})}\right\rangle$ and all the approximations, but here, for later use, we perform calculation for a general α; that is, we derive the position representation of $\left|j^{(1)}_{\kappa,\Delta,\alpha}\right\rangle$, $\left|j^{(2)}_{\gamma,\delta,\alpha}\right\rangle$, and $\left|j^{(3)}_{\beta,\alpha}\right\rangle$. We start with the derivation of Eq. (5.43). We have

$$\langle q|j^{(1)}_{\kappa,\Delta,\alpha}\rangle = \frac{1}{\sqrt{N^{(1)}_{\kappa,\Delta,j}}}\sum_{s\in\mathbb{Z}} e^{-\frac{1}{2}\kappa^2\alpha^2(ds+j)^2}\,\langle q|_{\hat{q}}\,X(\alpha(ds+j))S(-\ln\Delta)\,|0\rangle_f \tag{A.10}$$

$$= \sum_{s\in\mathbb{Z}} \frac{e^{-\frac{1}{2}\kappa^2\alpha^2(ds+j)^2}}{\sqrt{\Delta N^{(1)}_{\kappa,\Delta,j}}}\,\langle(q-\alpha(ds+j))/\Delta|0\rangle_f \tag{A.11}$$

$$= \left(\sqrt{\pi\Delta^2}N^{(1)}_{\kappa,\Delta,j}\right)^{-\frac{1}{2}}\sum_{s\in\mathbb{Z}} e^{-\frac{1}{2}\kappa^2\alpha^2d^2\left(s+\frac{j}{d}\right)^2-\frac{1}{2\Delta^2}\left(q-\alpha d\left(s+\frac{j}{d}\right)\right)^2} \tag{A.12}$$

$$= \left(\frac{2\sqrt{\pi\Delta^2}}{N^{(1)}_{\kappa,\Delta,j}}\right)^{\frac{1}{2}} E_{\frac{1}{\kappa^2},\alpha d,\frac{j}{d}} * G_{\Delta^2}(q), \tag{A.13}$$

where we used $\langle q|_{\hat{q}}\,X(a) = \langle q-a|_{\hat{q}}$ and $\langle q|_{\hat{q}}\,S(r) = \langle e^r q|_{\hat{q}}\,e^{r/2}$ in the second equality, and $\langle q|0\rangle_f = \pi^{-\frac{1}{4}}\exp(-q^2/2)$ in the third equality. Substituting α with α_d in Eq. (A.13), we obtain Eq. (5.43).

The derivation of Eq. (5.45) is similar. We have

$$\langle q|j^{(2)}_{\gamma,\delta,\alpha}\rangle = \frac{1}{\sqrt{N^{(2)}_{\gamma,\delta,j}}}\iint \frac{dr_1dr_2}{2\pi\gamma\delta}\, e^{-\frac{r_1^2}{2\gamma^2}-\frac{r_2^2}{2\delta^2}}\,\langle q|_{\hat{q}}\,V(\boldsymbol{r})\left|j^{(\mathrm{ideal})}\right\rangle \tag{A.14}$$

$$= \frac{1}{\sqrt{N^{(2)}_{\gamma,\delta,j}}}\iint \frac{dr_1dr_2}{2\pi\gamma\delta}\, e^{-\frac{r_1^2}{2\gamma^2}-\frac{r_2^2}{2\delta^2}-\frac{ir_1r_2}{2}+ir_1q}\left\langle q-r_2|j^{(\mathrm{ideal})}\right\rangle, \tag{A.15}$$

where we used $V(\boldsymbol{r}) := \exp(-ir_pr_q/2)Z(r_p)X(r_q)$, $\langle q|_{\hat{q}}\,Z(r_p) = \langle q|_{\hat{q}}\,e^{-ir_pq}$, and $\langle q|_{\hat{q}}\,X(r_q) = \left\langle q-r_q\right|_{\hat{q}}$. Using $\left\langle q-r_2|j^{(\mathrm{ideal})}\right\rangle = \sum_{s\in\mathbb{Z}}\delta\left(\alpha(ds+j)-q+r_2\right)$, we have

$$(A.15) = \left(\frac{\alpha d}{N^{(2)}_{\gamma,\delta,j}}\right)^{\frac{1}{2}}\iint \frac{dr_1dr_2}{2\pi\gamma\delta}\, e^{-\frac{r_1^2}{2\gamma^2}-\frac{r_2^2}{2\delta^2}-\frac{ir_1r_2}{2}+ir_1q}\sum_{s\in\mathbb{Z}}\delta\left(r_2-q+\alpha(ds+j)\right) \tag{A.16}$$

$$= \left(\frac{\alpha d}{N^{(2)}_{\gamma,\delta,j}}\right)^{\frac{1}{2}} \sum_{s\in\mathbb{Z}} \int \frac{dr_1}{2\pi\gamma\delta} e^{-\frac{1}{2\gamma^2}\left[r_1 - \frac{i\gamma^2}{2}(q+\alpha(ds+j))\right]^2 - \frac{\gamma^2}{8}(q+\alpha(ds+j))^2 - \frac{1}{2\delta^2}(q-\alpha(ds+j))^2} \tag{A.17}$$

$$= \left(\frac{\alpha d}{2\pi\delta^2 N^{(2)}_{\gamma,\delta,j}}\right)^{\frac{1}{2}} e^{-\frac{\lambda(\gamma,\delta)q^2}{2\delta^2}} \sum_{s\in\mathbb{Z}} e^{-\frac{\alpha^2 d^2 \lambda(\gamma,\delta)}{2\delta^2}\left(s+\frac{j}{d}\right)^2 + \frac{\alpha d q}{\delta^2}\left(\lambda(\gamma,\delta) - \frac{\gamma^2\delta^2}{2}\right)\left(s+\frac{j}{d}\right)} \tag{A.18}$$

$$= \left(\frac{\alpha d}{2\pi\delta^2 N^{(2)}_{\gamma,\delta,j}}\right)^{\frac{1}{2}} \sum_{s\in\mathbb{Z}} e^{-\frac{\lambda(\gamma,\delta)}{2\delta^2}\left[q - \alpha d\left(1-\frac{\gamma^2\delta^2}{2\lambda(\gamma,\delta)}\right)\left(s+\frac{j}{d}\right)\right]^2 - \frac{\alpha^2 d^2 \lambda(\gamma,\delta)}{2\delta^2}\left[1-\left(1-\frac{\gamma^2\delta^2}{2\lambda(\gamma,\delta)}\right)^2\right]\left(s+\frac{j}{d}\right)^2} \tag{A.19}$$

$$= \left(\frac{\alpha d}{\lambda(\gamma,\delta) N^{(2)}_{\gamma,\delta,j}}\right)^{\frac{1}{2}} E_{\frac{\lambda(\gamma,\delta)}{\gamma^2}\left(1-\frac{\gamma^2\delta^2}{2\lambda(\gamma,\delta)}\right)^2,\, \alpha d\left(1-\frac{\gamma^2\delta^2}{2\lambda(\gamma,\delta)}\right),\, \frac{j}{d}} * G_{\frac{\delta^2}{\lambda(\gamma,\delta)}}(q), \tag{A.20}$$

where we used a Gaussian integral in the third equality, and used

$$1 - \left(1 - \frac{\gamma^2\delta^2}{2\lambda(\gamma,\delta)}\right)^2 = \left(\frac{\gamma\delta}{\lambda(\gamma,\delta)}\right)^2 \tag{A.21}$$

in the last equality. Substituting α with α_d in Eq. (A.20) leads to Eq. (5.45).

The derivation of Eq. (5.47) needs a trick. We have

$$\langle q | j^{(3)}_{\beta,\alpha}\rangle = \frac{1}{\sqrt{N^{(3)}_{\beta,j}}} \langle q|_{\hat{q}} \sum_{n\in\mathbb{N}} |n\rangle\langle n|_f\, e^{-\beta\left(n+\frac{1}{2}\right)} \left|j^{(\mathrm{ideal})}\right\rangle \tag{A.22}$$

$$= \left(\frac{\alpha d}{N^{(3)}_{\beta,j}}\right)^{\frac{1}{2}} \sum_{s\in\mathbb{Z}} \sum_{n\in\mathbb{N}} e^{-\beta\left(n+\frac{1}{2}\right)} \psi_n(q)\psi_n^*(\alpha(ds+j)), \tag{A.23}$$

where $\psi_n(x) := (2^n n! \sqrt{\pi})^{-1/2} e^{-x^2/2} H_n(x)$ denotes the wave function of the Fock state. Using Mehler's Hermite polynomial formula [16]

$$\sum_{n\in\mathbb{N}} \frac{(u/2)^n}{n!} H_n(x) H_n(y) \exp\left(-\frac{x^2+y^2}{2}\right) = \frac{1}{\sqrt{1-u^2}} \exp\left[-\frac{(1+u^2)(x^2+y^2) - 4uxy}{2(1-u^2)}\right], \tag{A.24}$$

we obtain

$$\left\langle q | j_\beta^{(3)} \right\rangle = \left(\frac{\pi^{-1} e^{-\beta} \alpha d}{(1 - e^{-2\beta}) N_{\beta,j}^{(3)}} \right)^{\frac{1}{2}} \sum_{s \in \mathbb{Z}} e^{-\frac{(1+e^{-2\beta})\left(q^2 + \alpha^2 (ds+j)^2\right)}{2(1-e^{-2\beta})} - \frac{4e^{-\beta}\alpha(ds+j)q}{2(1-e^{-2\beta})}} \tag{A.25}$$

$$= \left(\frac{(2\pi)^{-1} \alpha d}{\sinh \beta N_{\beta,j}^{(3)}} \right)^{\frac{1}{2}} \sum_{s \in \mathbb{Z}} e^{-\frac{\alpha^2 d^2}{2\tanh\beta}\left(s + \frac{j}{d}\right)^2 + \frac{\alpha d q}{\sinh\beta}\left(s + \frac{j}{d}\right) - \frac{q^2}{2\tanh\beta}} \tag{A.26}$$

$$= \left(\frac{(2\pi)^{-1} \alpha d}{\sinh \beta N_{\beta,j}^{(3)}} \right)^{\frac{1}{2}} \sum_{s \in \mathbb{Z}} e^{-\frac{1}{2\tanh\beta}\left(q - \frac{\alpha d}{\cosh\beta}\left(s + \frac{j}{d}\right)\right)^2 - \frac{\alpha^2 d^2 \tanh\beta}{2}\left(s + \frac{j}{d}\right)^2} \tag{A.27}$$

$$= \left(\frac{\alpha d}{\cosh \beta \; N_{\beta,j}^{(3)}} \right)^{\frac{1}{2}} E_{\frac{1}{\sinh\beta\cosh\beta}, \frac{\alpha d}{\cosh\beta}, \frac{j}{d}} * G_{\tanh\beta}(q). \tag{A.28}$$

Substituting α with α_d in Eq. (A.28) leads to Eq. (5.47).

Next, we prove Lemma 5.3.3 to derive Eqs. (5.44), (5.46), and (5.48) from Eqs. (5.43), (5.45), and (5.47), respectively. From the definition of $E_{\mu,\Gamma,a}$ and $\tilde{E}_{\mu,\Gamma,a}$ in Definition 5.3.1 as well as the definition of G_ν in Eq. (5.40), we have

$$E_{\mu,\Gamma,a} * G_\nu(q) = \frac{1}{\sqrt{2\pi\nu}} \sum_{s \in \mathbb{Z}} \exp\left[-\frac{(s+a)^2 \Gamma^2}{2\mu} - \frac{(q - (s+a)\Gamma)^2}{2\nu} \right] \tag{A.29}$$

$$= \frac{e^{-\frac{1}{2\nu} q^2}}{\sqrt{2\pi\nu}} \vartheta\begin{bmatrix} a \\ 0 \end{bmatrix}\left(\frac{\Gamma q}{2\pi i \nu}, \frac{i(1 + \nu/\mu)\Gamma^2}{2\pi\nu} \right), \tag{A.30}$$

$$\tilde{E}_{\mu,\Gamma,a} * G_\nu(q) = \frac{1}{\sqrt{2\pi\nu}} \sum_{s \in \mathbb{Z}} \exp\left[-\frac{s^2 \Gamma^2}{2\mu} - \frac{(q + s\Gamma)^2}{2\nu} + 2\pi i a s \right] \tag{A.31}$$

$$= \frac{e^{-\frac{1}{2\nu} q^2}}{\sqrt{2\pi\nu}} \vartheta\begin{bmatrix} 0 \\ a \end{bmatrix}\left(\frac{i\Gamma q}{2\pi \nu}, \frac{i(1 + \nu/\mu)\Gamma^2}{2\pi\nu} \right). \tag{A.32}$$

The theta function has the following identity [12]

$$\vartheta(z/\tau, -1/\tau) = (-i\tau)^{\frac{1}{2}} \exp(\pi i z^2/\tau)\, \vartheta(z, \tau), \tag{A.33}$$

which leads to

$$\vartheta\begin{bmatrix} 0 \\ a \end{bmatrix}(z/\tau, -1/\tau) = (-i\tau)^{\frac{1}{2}} \exp(\pi i z^2/\tau)\, \vartheta\begin{bmatrix} a \\ 0 \end{bmatrix}(z, \tau). \tag{A.34}$$

Applying this to Eqs. (A.30) and (A.32), we have

$$(A.30) = \frac{e^{\left(-\frac{1}{2\nu}+\frac{1}{2\nu(1+\nu/\mu)}\right)q^2}}{\sqrt{(1+\nu/\mu)\Gamma^2}} \vartheta\begin{bmatrix}0\\a\end{bmatrix}\left(-\frac{q}{(1+\nu/\mu)\Gamma}, \frac{2\pi i\nu}{(1+\nu/\mu)\Gamma^2}\right) \tag{A.35}$$

$$= \sqrt{\frac{2\pi\mu}{\Gamma^2}}\, G_{\mu+\nu}(q)\, \vartheta\begin{bmatrix}0\\a\end{bmatrix}\left(-\frac{q}{(1+\nu/\mu)\Gamma}, \frac{2\pi i\nu}{(1+\nu/\mu)\Gamma^2}\right), \tag{A.36}$$

$$(A.32) = \sqrt{\frac{2\pi\mu}{\Gamma^2}}\, G_{\mu+\nu}(q)\, \vartheta\begin{bmatrix}a\\0\end{bmatrix}\left(-\frac{q}{(1+\nu/\mu)\Gamma}, \frac{2\pi i\nu}{(1+\nu/\mu)\Gamma^2}\right), \tag{A.37}$$

which proves Lemma 5.3.3. Then, as mentioned above, we obtain Eqs. (5.44), (5.46), and (5.48) by applying Lemma 5.3.3 to Eqs. (5.43), (5.45), and (5.47), respectively. □

A.2.2 *Proof of Proposition 5.3.10*

We compute $W_{|j_{\sigma_q^2,\sigma_p^2,\Gamma}\rangle\langle j'_{\sigma_q^2,\sigma_p^2,\Gamma}|}$ as follows:

$$W_{|j_{\sigma_q^2,\sigma_p^2,\Gamma}\rangle\langle j'_{\sigma_q^2,\sigma_p^2,\Gamma}|} = \frac{1}{\pi}\int dx\, e^{2ipx}\langle q-x|j_{\sigma_q^2,\sigma_p^2,\Gamma}\rangle\langle j'_{\sigma_q^2,\sigma_p^2,\Gamma}|q+x\rangle \tag{A.38}$$

$$= \frac{2\Gamma\left(\Lambda(\sigma_q^2,\sigma_p^2)\right)^{-\frac{1}{2}}}{\pi\sqrt{N_{\sigma_q^2,\sigma_p^2,\Gamma,j}N_{\sigma_q^2,\sigma_p^2,\Gamma,j'}}}\int dx\, e^{2ipx}\left(E_{\frac{\Lambda(\sigma_q^2,\sigma_p^2)}{2\sigma_p^2},\Gamma,\frac{j}{d}} * G_{2\sigma_q^2}(q-x)\right) \times\left(E_{\frac{\Lambda(\sigma_q^2,\sigma_p^2)}{2\sigma_p^2},\Gamma,\frac{j'}{d}} * G_{2\sigma_q^2}(q+x)\right) \tag{A.39}$$

$$= \frac{\left(2\pi^2\sigma_q^2\sqrt{\Lambda(\sigma_q^2,\sigma_p^2)}\right)^{-1}\Gamma}{\sqrt{N_{\sigma_q^2,\sigma_p^2,\Gamma,j}N_{\sigma_q^2,\sigma_p^2,\Gamma,j'}}}\int dx\, e^{2ipx}\sum_s \exp\left[-\frac{\left(s+\frac{j}{d}\right)^2\Gamma^2\sigma_p^2}{\Lambda(\sigma_q^2,\sigma_p^2)} - \frac{1}{4\sigma_q^2}\left(q-x-\left(s+\frac{j}{d}\right)\Gamma\right)^2\right] \times\sum_{s'}\exp\left[-\frac{\left(s'+\frac{j'}{d}\right)^2\Gamma^2\sigma_p^2}{\Lambda(\sigma_q^2,\sigma_p^2)} - \frac{1}{4\sigma_q^2}\left(q+x-\left(s'+\frac{j'}{d}\right)\Gamma\right)^2\right] \tag{A.40}$$

$$
\begin{aligned}
&= \frac{\left(2\pi^2\sigma_q^2\sqrt{\Lambda(\sigma_q^2,\sigma_p^2)}\right)^{-1}\Gamma}{\sqrt{N_{\sigma_q^2,\sigma_p^2,\Gamma,j}N_{\sigma_q^2,\sigma_p^2,\Gamma,j'}}} \int dx \sum_{s,s'} \exp\left(-\frac{1}{2\sigma_q^2}\left\{x - i\left[2\sigma_q^2 p + \frac{i\Gamma}{2}\left(s + \frac{j}{d} - s' - \frac{j'}{d}\right)\right]\right\}^2\right) \\
&\quad \times \exp\left\{-\frac{1}{2\sigma_q^2}\left[2\sigma_q^2 p + \frac{i\Gamma}{2}\left(s + \frac{j}{d} - s' - \frac{j'}{d}\right)\right]^2 - \frac{1}{2\sigma_q^2}\left[q^2 - \Gamma q\left(s + \frac{j}{d} + s' + \frac{j'}{d}\right)\right]\right\} \\
&\quad \times \exp\left\{-\frac{\Gamma^2}{2}\left(\frac{\sigma_p^2}{\Lambda(\sigma_q^2,\sigma_p^2)} + \frac{1}{4\sigma_q^2}\right)\left[\left(s + \frac{j}{d} + s' + \frac{j'}{d}\right)^2 + \left(s + \frac{j}{d} - s' - \frac{j'}{d}\right)^2\right]\right\}
\end{aligned} \tag{A.41}
$$

$$
\begin{aligned}
&= \frac{\left(2\pi^3\sigma_q^2\Lambda(\sigma_q^2,\sigma_p^2)\right)^{-\frac{1}{2}}\Gamma}{\sqrt{N_{\sigma_q^2,\sigma_p^2,\Gamma,j}N_{\sigma_q^2,\sigma_p^2,\Gamma,j'}}} \sum_{s,s'} \exp\left\{-\frac{\Gamma^2\sigma_p^2}{2\Lambda(\sigma_q^2,\sigma_p^2)}\left[\left(s + \frac{j}{d} + s' + \frac{j'}{d}\right)^2 + \left(s + \frac{j}{d} - s' - \frac{j'}{d}\right)^2\right]\right\} \\
&\quad \times \exp\left\{-\frac{1}{2\sigma_q^2}\left[q - \frac{\Gamma}{2}\left(s + \frac{j}{d} + s' + \frac{j'}{d}\right)\right]^2 - 2\sigma_q^2 p^2 - i\Gamma p\left(s + \frac{j}{d} - s' - \frac{j'}{d}\right)\right\}
\end{aligned} \tag{A.42}
$$

where we used the standard form (5.70) in the second equality. At this stage, we will change the variables for the summation from s and s' to $s + s'$ and $s - s'$. Since $s + s'$ and $s - s'$ have the same parity, the summation splits into two parts: one with $s + s' = 2t,\ s - s' = 2t',\ (t, t' \in \mathbb{Z})$ and the other with $s + s' = 2t + 1,\ s - s' = 2t' + 1$. Thus, we have

$$
\begin{aligned}
&(A.42) \\
&= \frac{\left(2\pi^3\sigma_q^2\Lambda(\sigma_q^2,\sigma_p^2)\right)^{-\frac{1}{2}}\Gamma}{\sqrt{N_{\sigma_q^2,\sigma_p^2,\Gamma,j}N_{\sigma_q^2,\sigma_p^2,\Gamma,j'}}} \sum_{t,t'}\left(\exp\left\{-\frac{1}{2\sigma_q^2}\left[q - \Gamma\left(t + \frac{j+j'}{2d}\right)\right]^2 - 2\sigma_q^2 p^2 - 2i\Gamma p\left(t' + \frac{j-j'}{2d}\right)\right\}\right. \\
&\qquad \times \exp\left\{-\frac{2\Gamma^2\sigma_p^2}{\Lambda(\sigma_q^2,\sigma_p^2)}\left[\left(t + \frac{j+j'}{2d}\right)^2 + \left(t' + \frac{j-j'}{2d}\right)^2\right]\right\} \\
&\qquad + \exp\left\{-\frac{1}{2\sigma_q^2}\left[q - \Gamma\left(t + \frac{j+j'}{2d} + \frac{1}{2}\right)\right]^2 - 2\sigma_q^2 p^2 - 2i\Gamma p\left(t' + \frac{j-j'}{2d} + \frac{1}{2}\right)\right\} \\
&\qquad \left. \times \exp\left\{-\frac{2\Gamma^2\sigma_p^2}{\Lambda(\sigma_q^2,\sigma_p^2)}\left[\left(t + \frac{j+j'}{2d} + \frac{1}{2}\right)^2 + \left(t' + \frac{j-j'}{2d} + \frac{1}{2}\right)^2\right]\right\}\right) \\
&= \frac{\left(\pi^2\Lambda(\sigma_q^2,\sigma_p^2)\right)^{-\frac{1}{2}}\Gamma}{\sqrt{N_{\sigma_q^2,\sigma_p^2,\Gamma,j}N_{\sigma_q^2,\sigma_p^2,\Gamma,j'}}} \left\{\left(E_{\frac{\Lambda(\sigma_q^2,\sigma_p^2)}{4\sigma_p^2},\Gamma,\frac{j+j'}{2d}} * G_{\sigma_q^2}(q)\right) e^{-2\sigma_q^2 p^2}\, \vartheta\begin{bmatrix}\frac{j-j'}{2d} \\ 0\end{bmatrix}\left(-\frac{\Gamma p}{\pi}, \frac{2i\Gamma^2\sigma_p^2}{\pi\Lambda(\sigma_q^2,\sigma_p^2)}\right)\right. \\
&\qquad \left. + \left(E_{\frac{\Lambda(\sigma_q^2,\sigma_p^2)}{4\sigma_p^2},\Gamma,\frac{j+j'}{2d}+\frac{1}{2}} * G_{\sigma_q^2}(q)\right) e^{-2\sigma_q^2 p^2}\, \vartheta\begin{bmatrix}\frac{j-j'}{2d} + \frac{1}{2} \\ 0\end{bmatrix}\left(-\frac{\Gamma p}{\pi}, \frac{2i\Gamma^2\sigma_p^2}{\pi\Lambda(\sigma_q^2,\sigma_p^2)}\right)\right\}
\end{aligned} \tag{A.43}
$$

$$
\begin{aligned}
&= \frac{\left(2\pi\sigma_p^2\right)^{-\frac{1}{2}}}{\sqrt{N_{\sigma_q^2,\sigma_p^2,\Gamma,j}N_{\sigma_q^2,\sigma_p^2,\Gamma,j'}}} \left\{\left(E_{\frac{\Lambda(\sigma_q^2,\sigma_p^2)}{4\sigma_p^2},\Gamma,\frac{j+j'}{2d}} * G_{\sigma_q^2}(q)\right) e^{-\frac{p^2}{2\sigma_p^2}}\, \vartheta\begin{bmatrix}0 \\ \frac{j-j'}{2d}\end{bmatrix}\left(\frac{ip\Lambda(\sigma_q^2,\sigma_p^2)}{2\Gamma\sigma_p^2}, \frac{\pi i\Lambda(\sigma_q^2,\sigma_p^2)}{2\Gamma^2\sigma_p^2}\right)\right. \\
&\qquad \left. + \left(E_{\frac{\Lambda(\sigma_q^2,\sigma_p^2)}{4\sigma_p^2},\Gamma,\frac{j+j'}{2d}+\frac{1}{2}} * G_{\sigma_q^2}(q)\right) e^{-\frac{p^2}{2\sigma_p^2}}\, \vartheta\begin{bmatrix}0 \\ \frac{j-j'}{2d} + \frac{1}{2}\end{bmatrix}\left(\frac{ip\Lambda(\sigma_q^2,\sigma_p^2)}{2\Gamma\sigma_q^2}, \frac{\pi i\Lambda(\sigma_q^2,\sigma_p^2)}{2\Gamma^2\sigma_p^2}\right)\right\}
\end{aligned} \tag{A.44}
$$

$$
\begin{aligned}
&= \frac{(2\pi\sigma_p^2)^{-\frac{1}{2}}}{\sqrt{N_{\sigma_q^2,\sigma_p^2,\Gamma,j} N_{\sigma_q^2,\sigma_p^2,\Gamma,j'}}} \\
&\quad \times \sum_t \Bigg\{ \left(E_{\frac{\Lambda(\sigma_q^2,\sigma_p^2)}{4\sigma_p^2},\Gamma,\frac{j+j'}{2d}} * G_{\sigma_q^2}(q) \right) e^{2\pi i t \frac{j-j'}{2d}} e^{-\frac{1}{2\sigma_p^2}\left(p+\frac{\pi t \Lambda(\sigma_q^2,\sigma_p^2)}{\Gamma}\right)^2 - \frac{2\pi^2 t^2 \sigma_q^2 \Lambda(\sigma_q^2,\sigma_p^2)}{\Gamma^2}} \\
&\qquad + \left(E_{\frac{\Lambda(\sigma_q^2,\sigma_p^2)}{4\sigma_p^2},\Gamma,\frac{j+j'}{2d}+\frac{1}{2}} * G_{\sigma_q^2}(q) \right) e^{2\pi i t\left(\frac{j-j'}{2d}+\frac{1}{2}\right)} e^{-\frac{1}{2\sigma_p^2}\left(p+\frac{\pi t \Lambda(\sigma_q^2,\sigma_p^2)}{\Gamma}\right)^2 - \frac{2\pi^2 t^2 \sigma_q^2 \Lambda(\sigma_q^2,\sigma_p^2)}{\Gamma^2}} \Bigg\}
\end{aligned} \tag{A.45}
$$

$$
\begin{aligned}
&= \frac{1}{\sqrt{N_{\sigma_q^2,\sigma_p^2,\Gamma,j} N_{\sigma_q^2,\sigma_p^2,\Gamma,j'}}} \Bigg[\left(E_{\frac{\Lambda(\sigma_q^2,\sigma_p^2)}{4\sigma_p^2},\Gamma,\frac{j+j'}{2d}} * G_{\sigma_q^2}(q) \right) \left(\tilde{E}_{\frac{\Lambda(\sigma_q^2,\sigma_p^2)}{4\sigma_q^2},\frac{\pi\Lambda(\sigma_q^2,\sigma_p^2)}{\Gamma},\frac{j-j'}{2d}} * G_{\sigma_p^2}(p) \right) \\
&\qquad + \left(E_{\frac{\Lambda(\sigma_q^2,\sigma_p^2)}{4\sigma_p^2},\Gamma,\frac{j+j'}{2d}+\frac{1}{2}} * G_{\sigma_q^2}(q) \right) \left(\tilde{E}_{\frac{\Lambda(\sigma_q^2,\sigma_p^2)}{4\sigma_q^2},\frac{\pi\Lambda(\sigma_q^2,\sigma_p^2)}{\Gamma},\frac{j-j'}{2d}+\frac{1}{2}} * G_{\sigma_p^2}(p) \right) \Bigg],
\end{aligned} \tag{A.46}
$$

where we used Eq. (A.34) in the third equality. □

A.2.3 *The Proof of Proposition 5.3.12*

In order to derive the average photon number of the approximate code state $\left|j_{\sigma_q^2,\sigma_p^2,\Gamma}\right\rangle$ in Definition 5.3.9, we first calculate the expectation values $\langle \hat{q}^2 \rangle_{|j_{\sigma_q^2,\sigma_p^2,\Gamma}\rangle}$ and $\langle \hat{p}^2 \rangle_{|j_{\sigma_q^2,\sigma_p^2,\Gamma}\rangle}$ of the squares $\hat{q}^2$ and $\hat{p}^2$ of the quadrature operators with respect to $\left|j_{\sigma_q^2,\sigma_p^2,\Gamma}\right\rangle$, using its Wigner function (5.79). Then, one can obtain the average photon number $\langle \hat{n} \rangle_{|j_{\sigma_q^2,\sigma_p^2,\Gamma}\rangle}$ of the state $\left|j_{\sigma_q^2,\sigma_p^2,\Gamma}\right\rangle$ by exploiting the fact that $\langle \hat{q}^2 + \hat{p}^2 \rangle_{|j_{\sigma_q^2,\sigma_p^2,\Gamma}\rangle} = \langle 2\hat{n} + 1 \rangle_{|j_{\sigma_q^2,\sigma_p^2,\Gamma}\rangle}$. We frequently use Eqs. (5.83), (5.84), (5.85), and (5.86) in the following calculation. Let $\mathrm{pr}(q)$ and $\widetilde{\mathrm{pr}}(p)$ be the probability density functions to obtain the values q and p in the $\hat{q}$- and $\hat{p}$-quadrature measurements, respectively. Then, they can be given by

$$
\mathrm{pr}(q) = \int dp\, W_{|j_{\sigma_q^2,\sigma_p^2,\Gamma}\rangle\langle j_{\sigma_q^2,\sigma_p^2,\Gamma}|}(q,p) \tag{A.47}
$$

$$
= \frac{1}{N_{\sigma_q^2,\sigma_p^2,\Gamma,j}} \left[c_1 E_{\frac{\Lambda(\sigma_q^2,\sigma_p^2)}{4\sigma_p^2},\Gamma,\frac{j}{d}} + c_2 E_{\frac{\Lambda(\sigma_q^2,\sigma_p^2)}{4\sigma_p^2},\Gamma,\frac{j}{d}+\frac{1}{2}} \right] * G_{\sigma_q^2}(q), \tag{A.48}
$$

$$
\widetilde{\mathrm{pr}}(p) = \int dq\, W_{\left|j_{\sigma_q^2,\sigma_p^2,\Gamma}\right\rangle\left\langle j_{\sigma_q^2,\sigma_p^2,\Gamma}\right|}(q,p) \tag{A.49}
$$

$$= \frac{1}{N_{\sigma_q^2,\sigma_p^2,\Gamma,j}} \left[c_3 \tilde{E}_{\frac{\Lambda(\sigma_q^2,\sigma_p^2)}{4\sigma_q^2}, \frac{\pi\Lambda(\sigma_q^2,\sigma_p^2)}{\Gamma}, 0} + c_4 \tilde{E}_{\frac{\Lambda(\sigma_q^2,\sigma_p^2)}{4\sigma_q^2}, \frac{\pi\Lambda(\sigma_q^2,\sigma_p^2)}{\Gamma}, \frac{1}{2}} \right] * G_{\sigma_p^2}(p), \tag{A.50}$$

where c_1, c_2, c_3, and c_4 are defined as

$$c_1 := \vartheta\begin{bmatrix}0\\0\end{bmatrix}\left(0, 2\pi i \Gamma^{-2}\sigma_q^2 \Lambda(\sigma_q^2, \sigma_p^2)\right), \tag{A.51}$$

$$c_2 := \vartheta\begin{bmatrix}0\\\frac{1}{2}\end{bmatrix}\left(0, 2\pi i \Gamma^{-2}\sigma_q^2 \Lambda(\sigma_q^2, \sigma_p^2)\right), \tag{A.52}$$

$$c_3 := \vartheta\begin{bmatrix}\frac{j}{d}\\0\end{bmatrix}\left(0, 2\pi^{-1} i \Gamma^{2}\sigma_p^2 \left[\Lambda(\sigma_q^2, \sigma_p^2)\right]^{-1}\right), \tag{A.53}$$

$$c_4 := \vartheta\begin{bmatrix}\frac{j}{d}+\frac{1}{2}\\0\end{bmatrix}\left(0, 2\pi^{-1} i \Gamma^{2}\sigma_p^2 \left[\Lambda(\sigma_q^2, \sigma_p^2)\right]^{-1}\right). \tag{A.54}$$

Note that the normalization constant $N_{\sigma_q^2,\sigma_p^2,\Gamma,j}$ satisfies $N_{\sigma_q^2,\sigma_p^2,\Gamma,j} = c_1c_3 + c_2c_4$ as shown in Eq. (5.80). Using $\mathrm{pr}(q)$, we calculate the expectation value of $\hat{q}^2$ as follows:

$$\langle \hat{q}^2 \rangle_{|j_{\sigma_q^2,\sigma_p^2,\Gamma}\rangle}$$

$$= \int dq\, q^2\, \mathrm{pr}(q) \tag{A.55}$$

$$= \int dq \int dr \frac{q^2}{N_{\sigma_q^2,\sigma_p^2,\Gamma,j}} \left[c_1 E_{\frac{\Lambda(\sigma_q^2,\sigma_p^2)}{4\sigma_p^2},\Gamma,\frac{j}{d}}(r) + c_2 E_{\frac{\Lambda(\sigma_q^2,\sigma_p^2)}{4\sigma_p^2},\Gamma,\frac{j}{d}+\frac{1}{2}}(r) \right] G_{\sigma_q^2}(q-r) \tag{A.56}$$

$$= \int dq \int dr \frac{r^2+2r(q-r)+(q-r)^2}{N_{\sigma_q^2,\sigma_p^2,\Gamma,j}} \left[c_1 E_{\frac{\Lambda(\sigma_q^2,\sigma_p^2)}{4\sigma_p^2},\Gamma,\frac{j}{d}}(r) + c_2 E_{\frac{\Lambda(\sigma_q^2,\sigma_p^2)}{4\sigma_p^2},\Gamma,\frac{j}{d}+\frac{1}{2}}(r) \right] G_{\sigma_q^2}(q-r) \tag{A.57}$$

$$= \int dq'\, q'^2 G_{\sigma_q^2}(q') + \int dr \frac{r^2}{N_{\sigma_q^2,\sigma_p^2,\Gamma,j}} \left[c_1 E_{\frac{\Lambda(\sigma_q^2,\sigma_p^2)}{4\sigma_p^2},\Gamma,\frac{j}{d}}(r) + c_2 E_{\frac{\Lambda(\sigma_q^2,\sigma_p^2)}{4\sigma_p^2},\Gamma,\frac{j}{d}+\frac{1}{2}}(r) \right] \tag{A.58}$$

$$= \sigma_q^2 + \int \frac{dr}{N_{\sigma_q^2,\sigma_p^2,\Gamma,j}} \left\{ c_1 \left[\sum_{s\in\mathbb{Z}} \Gamma^2 \left(s+\frac{j}{d}\right)^2 e^{-\frac{\Gamma^2}{2\mu}\left(s+\frac{j}{d}\right)^2} \delta\left(r - \Gamma\left(s+\frac{j}{d}\right)\right) \right] \right.$$

$$+ c_2 \left[\sum_{s\in\mathbb{Z}} \Gamma^2 \left(s + \frac{j}{d} + \frac{1}{2}\right)^2 e^{-\frac{\Gamma^2}{2\mu}\left(s+\frac{j}{d}+\frac{1}{2}\right)^2} \delta\left(r - \Gamma\left(s + \frac{j}{d} + \frac{1}{2}\right)\right)\right]\Bigg\}\Bigg|_{\mu=\frac{\Lambda(\sigma_q^2,\sigma_p^2)}{4\sigma_p^2}} \tag{A.59}$$

$$= \sigma_q^2 - \frac{2}{N_{\sigma_q^2,\sigma_p^2,\Gamma,j}} \frac{\partial}{\partial(\mu^{-1})} \left[c_1 \vartheta\begin{bmatrix} \frac{j}{d} \\ 0 \end{bmatrix}\left(0, \frac{i\Gamma^2}{2\pi\mu}\right) + c_2 \vartheta\begin{bmatrix} \frac{j}{d}+\frac{1}{2} \\ 0 \end{bmatrix}\left(0, \frac{i\Gamma^2}{2\pi\mu}\right)\right]\Bigg|_{\mu=\frac{\Lambda(\sigma_q^2,\sigma_p^2)}{4\sigma_p^2}}, \tag{A.60}$$

where we used the fact that $G_{\sigma_q^2}(x)$ has zero mean in the fourth and the fifth equality. In the same way, for the expectation value of $\hat{p}^2$, we have

$$\left\langle \hat{p}^2 \right\rangle_{|j_{\sigma_q^2,\sigma_p^2,\Gamma}\rangle} \tag{A.61}$$

$$= \sigma_p^2 - \frac{2}{N_{\sigma_q^2,\sigma_p^2,\Gamma,j}} \frac{\partial}{\partial(\mu'^{-1})} \left[c_3 \vartheta\begin{bmatrix} 0 \\ 0 \end{bmatrix}\left(0, \frac{\pi i[\Lambda(\sigma_q^2,\sigma_p^2)]^2}{2\mu'\Gamma^2}\right) + c_4 \vartheta\begin{bmatrix} 0 \\ \frac{1}{2} \end{bmatrix}\left(0, \frac{\pi i[\Lambda(\sigma_q^2,\sigma_p^2)]^2}{2\mu'\Gamma^2}\right)\right]\Bigg|_{\mu'=\frac{\Lambda(\sigma_q^2,\sigma_p^2)}{4\sigma_q^2}}. \tag{A.62}$$

Now we define $\tilde{N}_{\sigma_q^2,\sigma_p^2,\Gamma,j}(x, y)$ as

$$\begin{aligned} \tilde{N}_{\sigma_q^2,\sigma_p^2,\Gamma,j}(x, y) := {}& \vartheta\begin{bmatrix} \frac{j}{d} \\ 0 \end{bmatrix}\left(0, \frac{i\Gamma^2}{2\pi}x\right) \vartheta\begin{bmatrix} 0 \\ 0 \end{bmatrix}\left(0, \frac{\pi i[\Lambda(\sigma_q^2,\sigma_p^2)]^2}{2\Gamma^2}y\right) \\ &+ \vartheta\begin{bmatrix} \frac{j}{d}+\frac{1}{2} \\ 0 \end{bmatrix}\left(0, \frac{i\Gamma^2}{2\pi}x\right) \vartheta\begin{bmatrix} 0 \\ \frac{1}{2} \end{bmatrix}\left(0, \frac{\pi i[\Lambda(\sigma_q^2,\sigma_p^2)]^2}{2\Gamma^2}y\right), \end{aligned} \tag{A.63}$$

where $N_{\sigma_q^2,\sigma_p^2,\Gamma,j} = \tilde{N}_{\sigma_q^2,\sigma_p^2,\Gamma,j}\left(4\sigma_p^2[\Lambda(\sigma_q^2,\sigma_p^2)]^{-1}, 4\sigma_q^2[\Lambda(\sigma_q^2,\sigma_p^2)]^{-1}\right)$. Then, the average photon number is given by

$$\begin{aligned} \langle n \rangle_{|j_{\sigma_q^2,\sigma_p^2,\Gamma}\rangle} &= \frac{\langle \hat{q}^2 + \hat{p}^2 \rangle_{|j_{\sigma_q^2,\sigma_p^2,\Gamma}\rangle} - 1}{2} \\ &= \frac{\sigma_q^2 + \sigma_p^2 - 1}{2} - \left(\frac{\partial}{\partial x} + \frac{\partial}{\partial y}\right) \ln \tilde{N}_{\sigma_q^2,\sigma_p^2,\Gamma,j}(x, y)\Bigg|_{x=4\sigma_p^2[\Lambda(\sigma_q^2,\sigma_p^2)]^{-1},\, y=4\sigma_q^2[\Lambda(\sigma_q^2,\sigma_p^2)]^{-1}}, \end{aligned} \tag{A.64}$$

which proves Eq. (5.91). □

A.3 Alternative Expressions for the Wigner Function, Inner Products, and Average Photon Number of the Approximate GKP Code

In this appendix, we derive alternative expressions for the Wigner function, inner products, and the average photon number of the standard form $\left|j_{\sigma_q^2,\sigma_p^2,\Gamma}\right\rangle$ in terms of multi-variable generalization of the theta function, the Riemann theta function (also called Siegel theta function) [12]. For a row vector $z \in \mathbb{C}^n$ and a matrix $\boldsymbol{\tau} \in \mathbb{C}^n \times \mathbb{C}^n$ with $\boldsymbol{\tau} = \boldsymbol{\tau}^\top$ and $\mathrm{Im}(\boldsymbol{\tau}) > 0$, the Riemann theta function $\Theta\left[\begin{smallmatrix} a \\ b \end{smallmatrix}\right](z, \boldsymbol{\tau})$, is defined as

$$\Theta\begin{bmatrix} a \\ b \end{bmatrix}(z, \boldsymbol{\tau}) := \sum_{s \in \mathbb{Z}^n} \exp\left[\pi i (\boldsymbol{s} + \boldsymbol{a})\boldsymbol{\tau}(\boldsymbol{s} + \boldsymbol{a})^\top + 2\pi i (z + \boldsymbol{b})(\boldsymbol{s} + \boldsymbol{a})^\top\right]. \tag{A.65}$$

We also define multivariate normal distribution $\bar{G}[\nu](\boldsymbol{x})$ as

$$\bar{G}[\nu](\boldsymbol{x}) := \frac{1}{\sqrt{2\pi \det(\nu)}} \exp\left(-\frac{1}{2}\boldsymbol{x}\nu^{-1}\boldsymbol{x}^\top\right). \tag{A.66}$$

Now we define a multi-variable function combining $E_{\mu,\Gamma,a}(x)$ and $\tilde{E}_{\mu,\Gamma,a}(x)$ as follows.

Definition A.3.1 For a symmetric 2×2 matrix $\boldsymbol{\mu}$ satisfying $\mathrm{Re}(\boldsymbol{\mu}) > 0$ and 2-dimensional row vectors $\boldsymbol{\Gamma}$, $\boldsymbol{a}$, and $\boldsymbol{b}$, let $\bar{E}\left[\boldsymbol{\mu}, \boldsymbol{\Gamma}, \boldsymbol{a}, \boldsymbol{b}\right](\boldsymbol{x})$ be defined as

$$\bar{E}\left[\boldsymbol{\mu}, \boldsymbol{\Gamma}, \boldsymbol{a}, \boldsymbol{b}\right](\boldsymbol{x}) := \exp\left(-\frac{1}{2}\boldsymbol{x}\boldsymbol{\mu}^{-1}\boldsymbol{x}^\top\right) \sum_{s \in \mathbb{Z}^2} e^{2\pi i \boldsymbol{b}(\boldsymbol{s}+\boldsymbol{a})^\top} \delta(\boldsymbol{x} - (\boldsymbol{s} + \boldsymbol{a}) \circ \boldsymbol{\Gamma}), \tag{A.67}$$

where $\circ$ denotes an Hadamard product $(A \circ B)_{ij} = (A)_{ij}(B)_{ij}$.

A.3.1 An Aternative Expression of the Wigner Function

Under Definition A.3.1, we have an alternative expression of the Wigner function (5.79), which is relatively concise.

Corollary A.3.1 (Alternative expression of the Wigner function) *The Wigner function given in Eq. (5.79) is alternatively represented as*

$$
\begin{aligned}
&W_{|j_{\sigma_q^2,\sigma_p^2,\Gamma}\rangle\langle j'_{\sigma_q^2,\sigma_p^2,\Gamma}|}(q,p)\\
&=\frac{1}{\sqrt{N_{\sigma_q^2,\sigma_p^2,\Gamma,j}N_{\sigma_q^2,\sigma_p^2,\Gamma,j'}}}\bar{E}\Big[\Lambda(\sigma_q^2,\sigma_p^2)\begin{pmatrix}4\sigma_p^2 & 2i\\ 2i & 4\sigma_q^2\end{pmatrix}^{-1},\big(\tfrac{\Gamma}{2},\tfrac{\pi\Lambda(\sigma_q^2,\sigma_p^2)}{\Gamma}\big),\big(\tfrac{j+j'}{d},0\big),\big(0,\tfrac{j'}{d}\big)\Big]\\
&\qquad\vec{*}\,\bar{G}\Big[\begin{pmatrix}\sigma_q^2 & 0\\ 0 & \sigma_p^2\end{pmatrix}\Big](q,p),
\end{aligned}
\tag{A.68}
$$

where $\bar{E}$ is defined in Definition A.3.1, and $\vec{*}$ denotes a convolution in a multivariate sense.

Proof Comparing Eqs. (5.79) and (A.68), it is sufficient to show the following equality.

$$
\begin{aligned}
&\Big(E_{\mu,\Gamma,a}*G_{\sigma_q^2}(q)\Big)\Big(\tilde{E}_{\mu',\Gamma',a'}*G_{\sigma_p^2}(p)\Big)+\Big(E_{\mu,\Gamma,a+\frac{1}{2}}*G_{\sigma_q^2}(q)\Big)\Big(\tilde{E}_{\mu',\Gamma',a'+\frac{1}{2}}*G_{\sigma_p^2}(p)\Big)\\
&=\bar{E}\Big[\begin{pmatrix}\mu^{-1} & \frac{2\pi i}{\Gamma\Gamma'}\\ \frac{2\pi i}{\Gamma\Gamma'} & \mu'^{-1}\end{pmatrix}^{-1},\big(\tfrac{\Gamma}{2},\Gamma'\big),(2a,0),(0,a-a')\Big]\,\vec{*}\,\bar{G}\Big[\begin{pmatrix}\sigma_q^2 & 0\\ 0 & \sigma_p^2\end{pmatrix}\Big](q,p).
\end{aligned}
\tag{A.69}
$$

This follows from the following rearrangement of the summation.

$$
\begin{aligned}
&\Big(E_{\mu,\Gamma,a}*G_{\sigma_q^2}(q)\Big)\Big(\tilde{E}_{\mu',\Gamma',a'}*G_{\sigma_p^2}(p)\Big)+\Big(E_{\mu,\Gamma,a+\frac{1}{2}}*G_{\sigma_q^2}(q)\Big)\Big(\tilde{E}_{\mu',\Gamma',a'+\frac{1}{2}}*G_{\sigma_p^2}(p)\Big)\\
&=\iint dxdy\sum_{s,s'}\Big[\exp\Big(-\frac{x^2}{2\mu}\Big)\delta(x-(s+a)\Gamma)G_{\sigma_q^2}(q-x)\\
&\qquad\times\exp\Big(-\frac{y^2}{2\mu'}+2\pi ia's'\Big)\delta(y+s'\Gamma')G_{\sigma_p^2}(p-y)\\
&\qquad+\exp\Big(-\frac{x^2}{2\mu}\Big)\delta\Big(x-\big(s+a+\frac{1}{2}\big)\Gamma\Big)G_{\sigma_q^2}(q-x)\\
&\qquad\times\exp\Big(-\frac{y^2}{2\mu'}+2\pi i\big(a'+\frac{1}{2}\big)s'\Big)\delta(y+s'\Gamma')G_{\sigma_p^2}(p-y)\Big]
\end{aligned}
\tag{A.70}
$$

$$
\begin{aligned}
&=\iint dxdy\sum_{s,s'}\Big[\exp\Big(-\frac{x^2}{2\mu}\Big)\delta\Big(x-(2s+2a)\frac{\Gamma}{2}\Big)G_{\sigma_q^2}(q-x)\\
&\qquad\times\exp\Big(-\frac{y^2}{2\mu'}-2\pi ia's'-\pi i(2s)s'\Big)\delta(y-s'\Gamma')G_{\sigma_p^2}(p-y)\\
&\qquad+\exp\Big(-\frac{x^2}{2\mu}\Big)\delta\Big(x-(2s+1+2a)\frac{\Gamma}{2}\Big)G_{\sigma_q^2}(q-x)\\
&\qquad\times\exp\Big(-\frac{y^2}{2\mu'}-2\pi ia's'-\pi i(2s+1)s'\Big)\delta(y-s'\Gamma')G_{\sigma_p^2}(p-y)\Big]
\end{aligned}
\tag{A.71}
$$

$$
\begin{aligned}
&=\iint dxdy\sum_{s',s''}\exp\Big[-\frac{x^2}{2\mu}-\frac{y^2}{2\mu'}-\pi is''s'-2\pi ia's'\Big]\\
&\qquad\times\delta\Big(x-(s''+2a)\frac{\Gamma}{2}\Big)\delta(y-s'\Gamma')\,\bar{G}\Big[\begin{pmatrix}\sigma_q^2 & 0\\ 0 & \sigma_p^2\end{pmatrix}\Big](q-x,p-y)
\end{aligned}
\tag{A.72}
$$

$$= \iint dxdy \sum_{s',s''} \exp\Big[-\frac{1}{2}(x,y)\begin{pmatrix}\mu^{-1} & \frac{2\pi i}{\Gamma\Gamma'}\\ \frac{2\pi i}{\Gamma\Gamma'} & \mu'^{-1}\end{pmatrix}(x,y)^{\top} + 2\pi i(a-a')s'\Big]$$

$$\times\, \delta\Big(x-(s''+2a)\frac{\Gamma}{2}\Big)\delta(y-s'\Gamma')\,\bar{G}\Big[\begin{pmatrix}\sigma_q^2 & 0\\ 0 & \sigma_p^2\end{pmatrix}\Big](q-x,p-y) \tag{A.73}$$

$$= \bar{E}\Big[\begin{pmatrix}\mu^{-1} & \frac{2\pi i}{\Gamma\Gamma'}\\ \frac{2\pi i}{\Gamma\Gamma'} & \mu'^{-1}\end{pmatrix}^{-1}, \big(\tfrac{\Gamma}{2},\Gamma'\big), (2a,0), (0,a-a')\Big] \bar{*}\, \bar{G}\Big[\begin{pmatrix}\sigma_q^2 & 0\\ 0 & \sigma_p^2\end{pmatrix}\Big](q,p). \tag{A.74}$$

□

Note that the right-hand side of Eq. (A.68) approaches the right-hand side of Eq. (5.32) as $\sigma_q^2, \sigma_p^2 \to 0$. Equation (A.68) fits a viewpoint that a state corresponding to the Wigner function $\bar{E}(q,p)$ is subject to Gaussian random displacement channel [1], since random displacement can be represented as a convolution in the Wigner function picture. This viewpoint is utilized in numerical simulations of error analyses using approximate GKP codes [2, 4–7]. It should be noted that an operator corresponding to $\bar{E}(q,p)$ with parameters chosen as in Eq. (A.68) is neither a density operator nor a limit of density operators. There is thus no contradiction with the observation that an approximate GKP state differs from an ideal code state subject to random displacement noise, as stated in the explanation below the definition of Approximation 2.

A.3.2 Alternative Expressions of Normalization Constant and Inner Product

The alternative expression of the Wigner function in Corollary A.3.1 leads to the following alternative concise expressions of the normalization constant and inner product with the Riemann theta function.

Corollary A.3.2 (Alternative expressions of normalization constant and inner product) *The normalization factor given in Eq. (5.80) is alternatively represented as*

$$N_{\sigma_q^2,\sigma_p^2,\Gamma,j} = \Theta\begin{bmatrix}\big(\frac{2j}{d},0\big)\\ \big(0,\frac{j}{d}\big)\end{bmatrix}\left(\mathbf{0}, \begin{pmatrix}\frac{i\sigma_p^2\Gamma^2}{2\pi\Lambda(\sigma_q^2,\sigma_p^2)} & -\frac{1}{2}\\ -\frac{1}{2} & \frac{2\pi i\sigma_q^2\Lambda(\sigma_q^2,\sigma_p^2)}{\Gamma^2}\end{pmatrix}\right). \tag{A.75}$$

Furthermore, the inner product given in Eq. (5.81) is alternatively represented as

$$\langle j'_{\sigma_q^2,\sigma_p^2,\Gamma}|j_{\sigma_q^2,\sigma_p^2,\Gamma}\rangle = \frac{1}{\sqrt{N_{\sigma_q^2,\sigma_p^2,\Gamma,j}N_{\sigma_q^2,\sigma_p^2,\Gamma,j'}}}\Theta\begin{bmatrix}\big(\frac{j+j'}{d},0\big)\\ \big(0,\frac{j'}{d}\big)\end{bmatrix}\left(\mathbf{0}, \begin{pmatrix}\frac{i\sigma_p^2\Gamma^2}{2\pi\Lambda(\sigma_q^2,\sigma_p^2)} & -\frac{1}{2}\\ -\frac{1}{2} & \frac{2\pi i\sigma_q^2\Lambda(\sigma_q^2,\sigma_p^2)}{\Gamma^2}\end{pmatrix}\right). \tag{A.76}$$

Proof We combine Eq. (5.82) with the followings:

$$\int dx\, f(x) \circledast g(x) = \int dx\, f(x) \int dy\, g(y), \tag{A.77}$$

$$\begin{aligned}\int dx\, \bar{E}[\mu, \Gamma, a, b](x) &= \sum_{s\in\mathbb{Z}^2} \exp\left[-\tfrac{1}{2}((s+a)\circ\Gamma)^\top \mu^{-1}((s+a)\circ\Gamma) + 2\pi i b(s+a)^\top\right] \\ &= \Theta\begin{bmatrix}a\\b\end{bmatrix}\left(0, \frac{i}{2\pi}\mu^{-1}\circ\begin{pmatrix}\Gamma_1^2 & \Gamma_1\Gamma_2\\ \Gamma_1\Gamma_2 & \Gamma_2^2\end{pmatrix}\right),\end{aligned} \tag{A.78}$$

$$\int dx\, \bar{G}[\nu](x) = 1. \tag{A.79}$$

□

A.3.3 Alternative Expression of Average Photon Number

The alternative expression of the normalization constant leads to the following alternative expression of the average photon number with the Riemann theta function.

Corollary A.3.3 (Alternative expression of average photon number) *The average photon number given in Eq. (5.91) is alternatively represented as*

$$\begin{aligned}\langle\hat{n}\rangle_{|j_{\sigma_q^2,\sigma_p^2,\Gamma}\rangle} &= \frac{\sigma_q^2+\sigma_p^2-1}{2} \\ &\quad - \left(\frac{\partial}{\partial(\mu^{-1})_{11}} + \frac{\partial}{\partial(\mu^{-1})_{22}}\right)\ln \check{N}_{\sigma_q^2,\sigma_p^2,\Gamma,j}(\mu^{-1})\Big|_{\mu^{-1}=\frac{1}{\Lambda(\sigma_q^2,\sigma_p^2)}\begin{pmatrix}4\sigma_p^2 & 2i\\ 2i & 4\sigma_q^2\end{pmatrix}},\end{aligned} \tag{A.80}$$

where $\check{N}_{\sigma_q^2,\sigma_p^2,\Gamma,j}(\mu^{-1})$ *is given by*

$$\check{N}_{\sigma_q^2,\sigma_p^2,\Gamma,j}(\mu^{-1}) := \Theta\begin{bmatrix}\left(\frac{2j}{d},0\right)\\ \left(0,\frac{j}{d}\right)\end{bmatrix}\left(0, \frac{i}{2\pi}\mu^{-1}\circ\begin{pmatrix}\frac{\Gamma^2}{4} & \frac{\pi\Lambda(\sigma_q^2,\sigma_p^2)}{2}\\ \frac{\pi\Lambda(\sigma_q^2,\sigma_p^2)}{2} & \frac{\pi^2[\Lambda(\sigma_q^2,\sigma_p^2)]^2}{\Gamma^2}\end{pmatrix}\right). \tag{A.81}$$

Proof We give a more intuitive proof rather than direct calculation given in Sect. A.2.3. We see that $\langle\hat{q}^2+\hat{p}^2\rangle_{|j_{\sigma_q^2,\sigma_p^2,\Gamma}\rangle}$ denotes a second moment of the (quasi)probability distribution $W_{|j_{\sigma_q^2,\sigma_p^2,\Gamma}\rangle\langle j_{\sigma_q^2,\sigma_p^2,\Gamma}|}$, which is a convolution of "normalized" $\bar{E}$ and $\bar{G}[\nu]$. Since $\bar{G}[\nu](x)$ has zero mean, the second moment of $W_{|j_{\sigma_q^2,\sigma_p^2,\Gamma}\rangle\langle j_{\sigma_q^2,\sigma_p^2,\Gamma}|}$ is a summation of the second moment of "normalized" $\bar{E}$ and that of $\bar{G}[\nu]$. (Eq. (A.58) also shows this fact.) The second moment of $\bar{G}[\nu]$ is simply given by $\sigma_q^2 + \sigma_p^2$. On the other hand, the second moment of normalization times $\bar{E}$ is given by

$$\frac{1}{N_{\sigma_q^2,\sigma_p^2,\Gamma,j}} \int dx \, \|x\|^2 \bar{E}\left[\Lambda(\sigma_q^2,\sigma_p^2)\begin{pmatrix} 4\sigma_p^2 2i & \\ 2i & 4\sigma_q^2 \end{pmatrix}^{-1}, (\tfrac{\Gamma}{2}, \tfrac{\pi\Lambda(\sigma_q^2,\sigma_p^2)}{\Gamma}), (\tfrac{2j}{d}, 0), (0, \tfrac{j}{d})\right](x) \tag{A.82}$$

$$= -\frac{2}{N_{\sigma_q^2,\sigma_p^2,\Gamma,j}} \int dx$$

$$\left(\frac{\partial}{\partial(\mu^{-1})_{11}} + \frac{\partial}{\partial(\mu^{-1})_{22}}\right) \bar{E}\left[\mu, (\tfrac{\Gamma}{2}, \tfrac{\pi\Lambda(\sigma_q^2,\sigma_p^2)}{\Gamma}), (\tfrac{2j}{d}, 0), (0, \tfrac{j}{d})\right](x)\Big|_{\mu=\Lambda(\sigma_q^2,\sigma_p^2)\begin{pmatrix} 4\sigma_p^2 & 2i \\ 2i & 4\sigma_q^2 \end{pmatrix}^{-1}} \tag{A.83}$$

$$= -2\left(\frac{\partial}{\partial(\mu^{-1})_{11}} + \frac{\partial}{\partial(\mu^{-1})_{22}}\right) \ln \check{N}_{\sigma_q^2,\sigma_p^2,\Gamma,j}(\mu^{-1})\Big|_{\mu^{-1}=\frac{1}{\Lambda(\sigma_q^2,\sigma_p^2)}\begin{pmatrix} 4\sigma_p^2 & 2i \\ 2i & 4\sigma_q^2 \end{pmatrix}}, \tag{A.84}$$

where we used Eq. (A.78) in the last equality. Combining these with the relation $\langle \hat{q}^2 + \hat{p}^2 \rangle_{|j_{\sigma_q^2,\sigma_p^2,\Gamma}\rangle} = \langle 2\hat{n} + 1 \rangle_{|j_{\sigma_q^2,\sigma_p^2,\Gamma}\rangle}$ proves the statement. □

References

1. Caruso F, Giovannetti V, Holevo AS (2006) One-mode bosonic Gaussian channels: a full weak-degradability classification. New J Phys 8(12):310–310. https://doi.org/10.1088/1367-2630/8/12/310
2. Menicucci NC (2014) Fault-tolerant measurement-based quantum computing with continuous-variable cluster states. Phys Rev Lett 112(12):120504
3. Ketterer A et al (2016) Quantum information processing in phase space: a modular variables approach. Phys Rev A 94(2):022325
4. Fukui K et al (2018) High-threshold fault-tolerant quantum computation with analog quantum error correction. Phys Rev X 8:021054. https://doi.org/10.1103/PhysRevX.8.021054
5. Vuillot C et al (2019) Quantum error correction with the toric Gottesman- Kitaev-Preskill code. Phys Rev A 99(3):032344
6. Wang Y (2019) Quantum error correction with the GKP code and concatenation with stabilizer codes. arXiv: 1908.00147 [quant-ph]
7. Noh K, Chamberland C (2020) Fault-tolerant bosonic quantum error correction with the surface-Gottesman-Kitaev-Preskill code. Phys Rev A 101(1):012316
8. Weigand DJ, Terhal BM (2018) Generating grid states from Schrödinger-cat states without postselection. Phys Rev A 97(2):022341
9. Galetti D, Marchiolli MA (1996) Discrete coherent states and probability distributions in finite-dimensional spaces. Ann Phys 249(2):454–480
10. Terhal BM, Weigand D (2016) Encoding a qubit into a cavity mode in circuit QED using phase estimation. Physl Rev A 93(1):012315

11. Duivenvoorden K, Terhal BM, Weigand D (2017) Single-mode displacement sensor. Phys Rev A 95(1):012305
12. Mumford D, Musili C (2007) Tata Lectures on Theta. I (Modern Birkhäuser classics). Birkhäuser Boston Incorporated
13. Zak J (1968) Dynamics of electrons in solids in external fields. Phys Rev 168(3):686
14. Janssen AJEM (1982) Bargmann transform, Zak transform, and coherent states. J Math Phys 23(5):720–731
15. Mumford D, Nori M, Norman P (2007) Tata lectures on theta III, vol. 43. Springer, Berlin
16. Weisstein EW (2019) Mehler's hermite polynomial formula. From MathWorld-A Wolfram Web Resource. Visited on 28/06/2019. http://mathworld.wolfram.com/MehlersHermitePolynomialFormula.html